Get Wired for Success

ADVANCE PRAISE

Get Wired for Success teaches you the very skills Rod developed to achieve his business success and flow. These are not "airy fairy" concepts, but concepts rooted in *science. Get Wired for Success* can help you develop your mindset to achieve your business goals.

Shane Barker, Business Consultant, Kaizen Strategies.

After a string of burning failures in my business, I had the feeling that everything I did was doomed to fail. Thanks to *Get Wired for Success*, I realized my problem was my mindset, and I have wired *my* brain for success. My new business venture is the most exciting yet. Rod's life and work have really made a profound and positive impact in my life.

Terry Proto, VR Executive, Founder, and CEO, Ubereal.

Get Wired for Success is a mind-opening, captivating, and life-changing experience. It is a unique program to transform your personal and business life. Rod's many great insights are highly recommended to anyone who genuinely wants to power-up their approach to life and business and realize their full potential.

Mark Cummings, Lawyer, Principal, Cummings Law.

Get Wired for Success will change your life. Rod's ability to connect with the emotions that many of us have suffered is a rare trait. I would highly recommend *Get Wired for Success* to anyone who wants to be genuinely happy and prosperous.

Aaron Parkinson, CEO, 7 Mile Marketing SEZC.

We are drowning in a flood of information but suffering from a drought of wisdom. Rod has written an easy and instructive read that will ease the wisdom drought.

Tony Hansen, Accountant, Senior Partner,
Business Advisory Services, Findex.

Wow, Wow, Wow! *Get Wired for Success* is just brilliant! Everything that Rod said was crushing him crushed me to the point of no return. I lost it all! Thank you, Rod, for giving me the inspiration, the tools, and the courage to make my business really work this time!

Belinda Green, Financial Administrator
and Small Business Owner.

I love the way that *Get Wired for Success* emphasizes neuroplasticity. The 12 Keys provide a step-by-step guide to shape your neural pathways for greater success in life. We all have a lot to learn from Rod's experiences.

Dr. Cathy Warburton, Veterinarian,
Well-Being Consultant and Coach.

Get Wired for Success is not only a compelling read, but it's real. Real in the sense that I could relate to everything Rod spoke about—his hardships and uncertainty. To hear first-hand from someone who has not only overcome his hurdles but smashed them out of the park fills me with the confidence I need to wire my brain for success.

Daniel Mallia, Video Director, 1 Minute Media.

We used *Get Wired for Success* to go from dreams to a clear business vision, and from vision to real results. Success! We are now living the dream. Thanks Rod!

Georgie Pearson, Owner, South Brew Café.

I could not put *Get Wired for Success* down. It opened an entirely new world of learning and application. By sharing this information with my team and clients, it will make their lives profoundly different and better.

Dr. Diederik Gelderman, BVSc, MVS, MT-NLP,
Veterinary Business Coach and Author.

GET WIRED FOR SUCCESS!

HOW TO WIRE YOUR BRAIN FOR SUCCESS IN BUSINESS AND LIFE WITH NEUROSCIENCE-MADE-EASY

DR. ROD IRWIN

NEW YORK

LONDON • NASHVILLE • MELBOURNE • VANCOUVER

Get Wired for Success

How to Wire Your Brain for Success in Business and Life with Neuroscience-made-easy

Published in New York, New York, by Morgan James Publishing. Morgan James is a trademark of Morgan James, LLC. www.MorganJamesPublishing.com

This Get Wired For Success book does not supersede or replace advice from a financial or healthcare professional. It is recommended that you seek advice from a suitably qualified financial or healthcare professional before acting on any information contained in or provided by this book. Full details of this disclaimer can be found at www.getwiredforsuccess.com

ISBN 9781631951145 paperback
ISBN 9781631951152 eBook
Library of Congress Control Number: 2020934699

Cover Design by:
Christopher Kirk
www.GFSstudio.com

Interior Design by:
Chris Treccani
www.3dogcreative.net

To Celia, my love,
for the countless roles you play in my life.

CONTENTS

ACKNOWLEDGMENTS

I would like to thank the thousands of neuroscientists and psychologists who have made this book possible. Most research work goes unnoticed, yet each generation of scientists builds on the previous one's key learnings until some of life's big questions can be answered, for example, "Can you wire your brain for the specific outcomes you desire?"

Thanks to their diligence and discipline, this question can now be answered unequivocally: "Yes, you *can* get wired for success!"

A big thank you to Shane Barker for his early encouragement and contributions.

My gratitude to Bruce Langdon for his creativity and elegant website design.

Thanks to my accountant, Tony Hansen at Findex, to Mark Cummings of Cummings Lawyers for all things legal, and to Dr. Lisa Lines' team at Capstone Editing, who worked my rough manuscript into a polished product with scientific precision.

When I submitted my manuscript to my publisher of choice, Morgan James, I did not expect a favorable reply within six hours and a signed contract within three weeks! Thank you for your faith and the support your team has given me.

Although I have never met them, I am also in debt to Brendon Burchard, who inspired me to make my message a multimedia presentation, and to Russell Brunson, who taught me how to turn science into a storyline.

INTRODUCTION

Business failure sucks, but so does mediocrity!

When I bought my business for over AUD$250,000, I soon discovered it was making a loss. In the next eight long years, I went AUD$1.1 million into debt. I was drained by poor profit, disengaged employees, and demanding clients. I suffered from crippling anxiety, mind-numbing insomnia, and even a near-death experience!

During that living nightmare, I often asked myself: Why are other people so successful? They have amazing purpose, personal power, and red-hot passion; they have wealth, freedom, and happiness; they seem *wired* for success—what is their secret?

Do *you* ever ask such questions? Do you wish you could create the business life of your dreams and live a life you *love*?

Here is the rub: The greater the gap between where you are now and your dreams of success, the greater your frustrations, negative thinking and emotions, and even the anxieties and fears that go with them—blocking the very success and happiness you so deeply desire, deserve, and dream of.

I get it. I have been there. I hated mediocrity and the constant, grinding fear of failure. Business success was not on my radar. I was in survival mode. My business nearly killed me!

Then, my big breakthrough: I discovered how to *wire my brain for success*—to think, feel, and see my world like successful people do.

And SNAP!

It *totally* changed my business. Now, I have a great bunch of happy clients, a highly motivated team, and a 721% jump in profits. I created the business life of my dreams, and it totally transformed me. I am now calm, confident, and living a life I *love*. It feels sensational, breathtaking, and indeed mind-blowing!

Here is an important point: World-leading neuroscientists and psychologists have proven that you can rewire your brain for specific outcomes. I am a scientist, and I applied the science with laser-like precision to my business life—WOW! What a difference!

What was the real secret to my success? After years of research and trial and error, I distilled a massive amount of science into just 12 easily understandable, essential facts that made the critical difference. I now call these facts "The 12 Keys to Wire Your Brain for Success."

Apply The 12 Keys, and it will be like your brain is turbocharged! Your thinking and emotions and the way you see your world will be so much better. Here is the best bit—the benefits you receive are lifelong; they are yours to keep forever.

So, imagine if you could wire *your* brain for success!

Imagine you could program *your* brain for __________. (Just fill in this space with the outcomes *you* desire!)

Imagine if, at some point in the future, you could look back at yourself reading these words and say, "Wow, what a sensational, mind-blowing, life-changing difference!"

Wouldn't *that* be amazing?

Well, now you can, and I can teach you how to do it. But wait! Before we get too far ahead of ourselves, if a little seed of doubt has just crept into your mind, let me guess what you might be thinking:

"Success sounds great, Rod, but I cannot change my brain. I am who I am!"

I used to think like that. We are totally unaware of how our brains work as we go about our daily lives, so we cannot imagine this seemingly impossible skill. I will explain by way of an analogy.

Your brain is like an extraordinary supercomputer with billions upon billions of connections; like a computer, little currents of electricity run between the connections to make it all work.

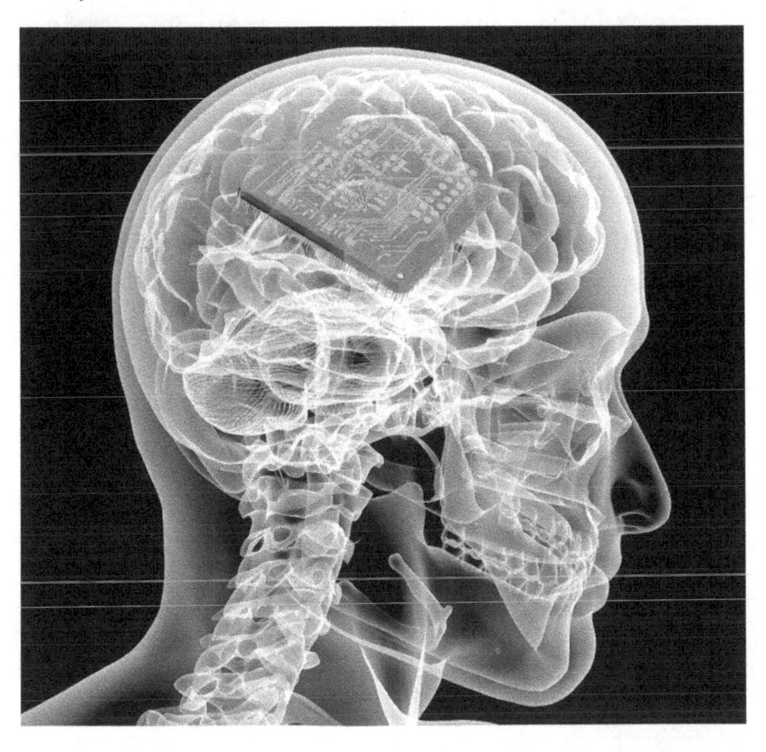

Just as you can upgrade computer chips or software, perhaps improving the processing power or speed, you can upgrade your strengths by wiring in better connections so that they become your *superpowers*.

Critically, just like you can replace defective chips or software, you can also learn to replace your negative thoughts and emotions, even your anxieties and fears, by rewiring the connections involved so that they become fading memories.

These two changes—upgrading with new wiring, and replacement by rewiring—enable a third change: They create capacity!

You can create extra capacity in a computer by inserting a new drive. When you delete your negative thought patterns and emotions, and their drain on your energy and time, you generate extra capacity in your mind for greater creativity, imagination, growth, contribution, compassion, clarity, gratitude, relaxation, happiness, love, and a host of other empowering states of mind.

How great would that be?

At this point, you might be silently screaming in frustration: "But Rod—my brain is not a computer! I cannot reach into my head and simply change the connections like you can change a computer chip or software."

True! So how did I wire my brain for success? I did it with *science*, specifically neuroscience and positive psychology; even more precisely, by applying this science to best business practices; and with even greater precision, by using The 12 Keys to Wire Your Brain for Success—trialed, refined, honed, and polished into a robust model that works under a wide variety of conditions and circumstances.

To be clear, this is not some new-age "believe it will be all right and trust in the future" quasi-religion in which happiness, freedom, or money fall from the sky just because you believe they will. The 12 Keys are *scientifically proven.*

Here is the next major point: Wiring your brain for success is a skill. It is easy to learn, and this is the real buzz: Once you get the hang of it, it becomes increasingly easier. You create progressively better results—so much better than missing out on wonderful opportunities, being mired in mediocrity, or, even worse, failing.

Now, with science, you can switch off that "autopilot feeling," in which you do not feel fully in control of your thoughts, emotions, or destiny. Instead, you can control your mind and not be controlled by it; you can direct your thoughts and emotions; and you can design your future and drive your destiny. Think of the possibilities!

This neatly brings us to a whole new set of questions that might be forming in your mind: "Okay, Rod… I think I might be able to wire my brain for success by using science, but why does that translate into creating the business life of my dreams? How do neuroscience and positive psychology improve my bottom line? Why do they boost profits?"

"Come to think of it, why does wiring my brain for success make me more *money*?"

Let me explain.

It was late one January, just a few days before I took over sole ownership of my business. It had not been performing well financially. I was one of three partners, and the existing partnership had been dysfunctional for some time. My partners wanted to retire, and I saw sole ownership as a great opportunity for improvement.

I knew what I wanted to create: a model business—my dream business. There was only one problem: I did not know how to do it.

I was standing in the office with the business manager at that time, and I can remember our conversation as if it were only yesterday: "So," I said. "How do you run the business?" (This question gives you some insight into how ill-prepared I was for the journey ahead.)

"It's all in there!" she said, pointing to a two-drawer, gray filing cabinet under the office desk.

My reaction was one of disappointment. "That's it?" I thought business management would be a lot more exciting.

"That's it," she replied. "Oh, and here's the wages book."

"Wages?" Suddenly, the responsibility of what I was about to take on began to weigh on me.

Just then, our receptionist, who had been listening to the conversation, piped up: "I can do the wages."

"Great," I said, at once relieved, but also with a nagging feeling that I had just let go of responsibility for something that was important and confidential.

As everyone else went about their work, I found myself alone in the office. I opened the top drawer of the gray filing cabinet and read the folder tab headings: Insurance, Yellow Pages, Claims, Licenses, Bank Statements, Adverts, etc. It seemed somewhat random but relatively straightforward. Was that it? Was that all there was to running a business?

Then, reality set in. How could I turn a dysfunctional business into a high-performance machine? I was about to take on a significant amount of debt to purchase it, so how could I turn its financial performance around? I did not want to sink!

Well, it would be fair to say that, for the next eight years, I did not sink—I swam extremely hard. One of our team members at the time said: "Rod, you're like a duck swimming upstream. Calm on the surface, but under the water, the legs are going flat out." She was right.

I made significant early progress, but four years after buying my business, I had to purchase the real estate as well, and my former business partners—now my landlords—drove a hard bargain.

My debt reached seven figures—well into seven figures!

Everything was maxed out: personal and business bank accounts, business loans, overdrafts, and credit cards. I moved money from one account to the next to pay off interest as it was due, desperately searching for extra cash like a rat searching for crumbs in a famine.

Everything was on the line—but for what? An ill-defined dream! The stress became appalling.

Managing a heavy workload in tough, adverse conditions, attracting and retaining clients in a very competitive environment, and trying but failing to retain staff *and* turn a profit were a massive load for one person who had no training as a manager, leader, owner, or entrepreneur. There was a jigsaw puzzle in my head, and I could not solve it as I was missing a great many pieces.

I realized I had created my very own Mount Everest, and my mountain was covered in mist. I had no route map, no GPS, no Sherpa, no supplementary oxygen, no experience, and no shelter. Disaster loomed.

The result? Low margins, poor profit, demanding clients, disengaged employees, distractions, doubts, stress, chronically low self-esteem, and insomnia—I was an accident waiting to happen.

It was not as though I was technically incompetent or stupid—I had a university degree with honors, followed by a master's degree in my chosen field. Yet, my Everest seemed to get steeper, colder, and more dangerous.

It was not as though I lived in a vacuum. I had read business management books, taken marketing courses, and had even gone to self-improvement workshops. I still lived in a world of pain.

It was not as though I was alone on this journey. Friends and family gave me well-meaning advice and pep talks. "Get a winning mindset," they said. I plunged on in survival mode.

My insomnia was truly awful—lying awake in bed night after endless night, thrashing around trying to find solutions to my problems, feeling an energy-sapping coldness inside me each day, struggling to concentrate, and battling to inspire and lead my team and serve my clients.

At my lowest point, I suffered from acute clinical anxiety, which is a completely different beast from the anxiety one might feel before an exam, when giving a presentation, or meeting someone important. Those who have suffered from it will have their own definitions of the condition. The following is mine.

On the way to visit clients in my car, I would be gripped by an awful pressure across my chest, so tight I could hardly breathe, and I would be forced to pull over and stop. Then, the panic would set in, my breaths coming in shallow, rapid gasps. In the driving mirror, my eyes would bulge with the fear of failure. I was out of control, adrenalin coursing through my veins!

It got worse. At 4.00 a.m., I would find myself lying on the cold tiles of my kitchen floor in the fetal position, crying in desperation, my limbs locked in muscle spasm, knowing that I had to go to work each morning, totally exhausted, but act as though everything was okay.

I had so many responsibilities—to my team, my clients, and my dear family. I could not let them down, but I was over $1 million in debt, and I did not know how to climb this mountain I had created.

The chronic stress began to destroy my health and well-being. One night, I found myself in hospital with an irregular heartbeat,[1] hooked up to dozens of leads, my wife standing by my bedside with tears rolling down her cheeks. How had it gotten to this point? Where was my life going? What had happened to my dreams? They had turned into a living nightmare.

Yes, my business very nearly killed me!

Little did I know that the seeds of this stress—this breakdown and failure—had been set long before I bought my business; they had been set in my childhood and adolescence and as I had grown into an adult. The seeds of failure were in my character, beliefs, and very psyche. They were set because I was human—no more, no less—with very human strengths and equally human weaknesses.[2]

These were just some of the weaknesses in my makeup when I purchased my business:

- I had a pessimistic view of the world, especially when it came to money. When presented with a poor set of financial

1 I later learned that increased levels of a hormone called cortisol are released into the bloodstream due to chronic stress, which in turn can cause irregular heartbeats.

2 Throughout *Get Wired for Success*, I illustrate how negative life experiences can set us up for mediocrity or failure by relating stories like the one above and how I conquered their devastating effects. You will have your own stories that have shaped your life, very negatively for some of you. Use the information in this book to rewrite your scripts!

results, I saw doom and gloom and not the opportunities that lay within the results.

Every time I saw a financial spreadsheet with a negative result, a shiver of fear ran through me, like ice water pouring down my back. It was the raw fear of loss—losing everything my wife and I had worked so hard for. My mindset was of scarcity and not of abundance; of meanness and not of generosity; of scrimping and not of investment.

- I also had a fear of interpersonal conflict. Whenever I had to confront one of my team or a client, it made me feel physically sick. I used to feel warm nausea settle in my stomach, and I could literally feel my heart thumping in my chest, so I totally avoided important issues. Big problems within my team or with my clients festered, and the business suffered.
- I had the mindset that if I were a strong leader, I would be rejected or disliked. This was fear of rejection. Growth, mutual understanding, and progress were blocked while flight to safety, a lack of communication, and stagnation flourished.
- Although I had dreams of a wonderful future, they lacked the clarity required for real and enduring success.

Decisions were made haphazardly, without real research, planning, or discussion. The big picture was missing, motivation was poor, and my personal power was weak.

Would you get into a plane knowing that the pilot has no compass, flight plan, or training, with the flight heading directly into a storm? That is the way I ran my business. As you now know, I crashed!

However, I also had significant strengths: good communication skills, genuine empathy for my team and clients, an eye for detail, and a sense of humor and fun—all desirable traits for a successful business life. I had blind faith that my strengths would see me through—they did not. I believed that if I were a good role model, my team would follow me—they did not. I trusted that my good intentions would be enough for my clients—they were not.

The pressure of owning and running a business exposed my weaknesses and crushed my strengths, and *because I did not realize this fact*, my business nightmare was the result.

At the time of my business purchase, I was totally unaware of the effect my personality would have on my success. I was just… well, me—just an ordinary guy who wanted to be a business owner and an entrepreneur, to make a difference to the world, and (above all) to provide well for his family.

So, to answer those important questions—how did neuroscience and positive psychology transform my business into the one of my dreams? How did they massively boost my bottom line? How did they turbocharge my business profits?

Yes, why does wiring your brain for success make you more money?

Let us return to the computer analogy. Imagine if my brain could be viewed as a supercomputer, and each one of my negative thought patterns and emotions were a defective or broken program or chip—one for financial pessimism, one for fear of conflict, one for a poorly defined business vision, and so on. Remember, I went more than $1 million into debt *and* my brain was *preloaded* with defective or broken thought patterns and emotions, just as surely as if they were faulty computer programs or chips.

I was hardwired for disaster and totally unaware of the fact.

And if my strengths could also be viewed as separate computer programs or chips, each one was in desperate need of an upgrade.

Then, I read a book introducing me to neuroscience. It totally changed my life.[3] This was my eureka moment! The light bulb went on—out of the dark and into the light! With a *lot* of research, and two to three years of trial and error, I learned how to use little-known branches of science to wire my brain for success.

How did I create this change, this dramatic turnaround, and this total transformation?

Back to the computer analogy:

- If my pessimism over money could be imagined as a silicon chip, I replaced it with one for optimism. Doing so not only turned the financial performance of my business around but also inspired my team!
- My fear of conflict? I installed a new software program for courage and quickly resolved the issues that had simmered in my business.
- If my ill-defined dreams for my business could be viewed as a computer graphic, I replaced them with high-definition, crystal-clear images of a wonderful and energizing future.

What about the chips or programs representing my strengths? They all received a *massive* upgrade, turning my strengths into superpowers! Communication, empathy, an eye for quality, and a sense of fun became key values in my business, uniting and motivating my team.

As previously mentioned, these two changes—replacement and upgrading—enabled a third change: They created capacity, freedom, and the realization of amazing potential.

You see, I always had this feeling deep within me that things could be so much better, that I had so much potential waiting to

---

3 See Chapter 1 for more information on this book.

be unleashed, and all I wanted to do was break out from everything holding me back—to burst free and achieve my true greatness. Now, I suddenly felt *unlocked* from my past, fully *engaged* with the present, and *unleashed* into an exciting future. Wow!

Fast-forward a few years: BOOM! Wiring my brain for success totally changed my business performance—efficiency, effectiveness, and energy all vastly improved.

Suddenly, I loved my work! I developed a great list of clients and a happy, loyal, and motivated team.

Importantly, running my business became easy. All the stresses of the past fell away. Now, I know exactly what is required. If I am away, it runs perfectly well by itself. My team works to a high level of expertise—knowing exactly what to do, how and when to do it, and, importantly, *why* they do it that way—and has a clear understanding of my high standards and expectations.

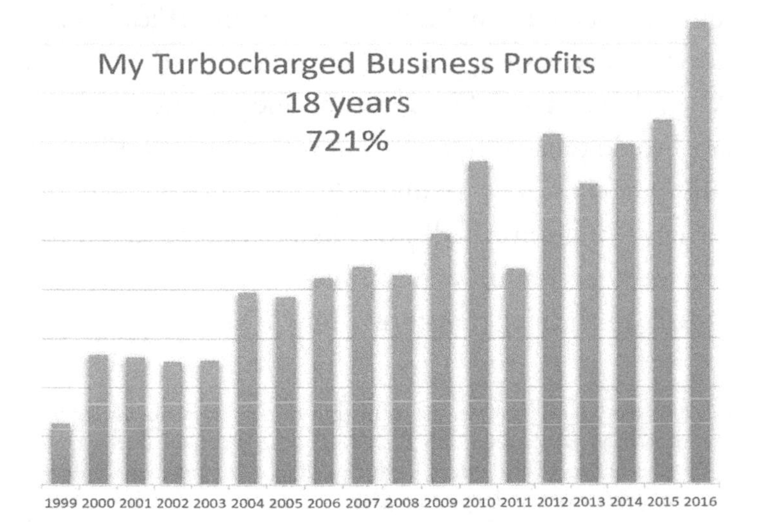

Profit soared! I took my business to a 721% increase in net profit. During this period, it powered through the Global Financial Crisis as if it had never happened.[4]

A *highly successful* business easily outperforms the stock market or real estate, hands down! I had wired my brain for success and transformed my business into a money-making machine. I had not just created the business I had imagined when I had first bought it—it was far, far better. I had created my *dream* business.

So, would you like to wire your brain for success and create the business life of your dreams? It sounds great, but another thought might have popped into your head:

"Yes, Rod, but it is not all about the money. What about my lifestyle?"

Another good question! Lifestyle issues are critical. Remember my chronic anxiety, stress, and irregular heartbeat? Not good.

When I wired my brain for business success, I suddenly realized I could transform my life, too! I rewired some broken and defective programs in my brain—like low self-esteem, self-doubt, and anxiety—and I installed some high-performance programs—like confidence, certainty, and calmness.

SNAP! An amazing transformation! What a difference!

4 The 2009 bar on the graph represents my business performance during the Australian financial year from July 1, 2008 to June 30, 2009. This was the period of the global financial crisis, during which Lehman Brothers failed and the Dow Jones Index hit its lowest level of 6,443.27 on March 6, 2009, from a high of 14,164.53 on October 9, 2007. See https://en.wikipedia.org/wiki/Financial_crisis_of_2007%E2%80%9308#Timeline.

Now, I really enjoy the rewards of my business success. My wife and I live in a dream home and have amazing international holidays, staying at the best hotels around the world.

Our family is united and strong. Our children are now making their own way in the world, and they love us as mentors and friends as well as parents.

Even more important than business success, my low self-esteem, awful insomnia, and acute clinical anxiety are now just distant memories, no longer controlling or blocking me. I cannot tell you how wonderful it is to write these words!

I became relaxed, confident, happy, and passionate. I have a deep inner calmness and contentment that comes from achieving a big vision, dreaming big again, and then creating new visions and adventures. I have no money worries, the freedom to do what I want, and, yes, real happiness.

Not only have I created my dream business, I now live a life I love—because I discovered how to wire my brain for success. I transformed my thinking and emotions and my view of my world. I transformed myself and, in turn, the results I got.

Imagine creating *your* version of a life *you* love! How great would that be?

Another great set of questions might be forming in your mind.

"Wiring my brain for success sounds cool, Rod, but maybe a little weird. Would I become a different person: less authentic or too greedy? What does it really *feel* like? How does it change *me*?"

I will explain with three analogies:

Analogy Number One: You know how you sometimes hang onto old technology for too long and then suddenly need a replacement—a new phone, a tablet or laptop, or perhaps a television? You bite the bullet and buy a new one. You power it up and "Wow!"

The power, logic, definition, color, and clarity are all amazing. *Everything feels better, lighter, smoother, faster, deeper, and richer.* That is what wiring your brain for success feels like. You are still the same authentic you, but your thinking, emotions, and the way you see your world is so much better.

Analogy Number Two: Sometimes, our lives are a daily grind. Imagine escaping that grind; imagine finding yourself on a movie set and *you* are the star of the movie! Like many Hollywood blockbusters, this movie has a gritty and difficult beginning, a challenging storyline with many ups and downs, and a wonderful ending in which, against the odds, you win.

Instead of being the star of an imaginary movie, you are *the star of the movie that is your real life! That* is what wiring your brain for success feels like.

Analogy Number Three: My wife has a brand new, red-hot Ford Mustang convertible. It looks amazing. On its dash is a simple rocker switch with the word "mode." Flick the switch to the "sport" mode, and everything changes via the onboard electronics—more responsive steering, quicker acceleration, tighter handling, and better braking—turbocharged! So much fun!

Like driving the Mustang, I can select the mode I want at the flick of a switch, and I have learned to do it quickly and easily. When I am feeling stressed or anxious, I can flick the switch to feel calmness and tranquility. Years ago, I used to think I was too weak for real business success, but now I can select the mode I want: courage, tenacity, or resilience. The way ahead used to be foggy and lack detail. Now, I view my world with crystal-clear clarity.

I am not just living the dream life—I am living a life *beyond* my wildest dreams.

Yes, it *is* sensational and breathtaking—even mind-blowing—and like driving the Mustang, *it is so much fun*.

So, if all this talk of success sounds like a distant dream to you, I really *do* know how that feels. You have read a little of my story—my Mount Everest, my world of debt, my life in survival mode, and my anxieties and fears.

Are you looking at your Mount Everest? Do you want to create the business life of your dreams and live a life you love? Let me ask you some key questions to find out if or how I can help you.

If I showed you *how* to wire your brain for success with proven science, and then I showed you *why* it would make you lots more money with much less stress and *what* it would feel like, do you think all this information would work for *you*?

You see, during all those hard years of research and trial and error, I suspected there would be other businesspeople who would say yes to that question. In fact, right from the "eureka moment," I sensed that other businesspeople would desire this amazing science and this incredible knowledge that promises to make a profound difference not only to their lives, but also the lives of the people who interact with them—their teams, clients, and community.

A quick Facebook poll confirmed my suspicions: People all over the world wanted to be wired for success, so *Get Wired for Success*—this book and the companion online course—was conceived, written, and produced!

The 12 Keys are now a robust, powerful, and easily understood model built for businesspeople to use every day in their busy, demanding, and sometimes chaotic business lives.

Critically, I have applied the Keys to other vital areas of business success—vision, creativity, motivation, emotional intelligence (EQ), leadership, and so on. They work seamlessly.

Have you ever had a feeling deep inside that things could be so much better, that you have so much potential waiting to be unleashed, and that there are limitations in your business life—some of which you can identify and some perhaps holding you back just below the surface of your consciousness? All you want is to break out from everything holding you back… to burst free and achieve your true greatness?

Yes? Then *Get Wired for Success* is perfect for you!

This book and the companion online course are carefully designed to take you from where you are now—your Point A—to a new point—your Point B—creating the business life of your dreams and living a life you love, as easily, quickly, and painlessly as possible so as to make your dreams reality.

It is not just about reaching targets in 30 or 10 years, next year, or even tomorrow. It is about making sure today is amazing… and then tomorrow, and then the day after that.

So, would you like to crush your anxieties and fears with powerful positive thinking and emotions so that they become just fading memories and you can transform your strengths into superpowers?

Do you want to generate extra capacity and time for greater creativity, imagination, growth, contribution, compassion, gratitude, relaxation, happiness, love, and a host of other empowering states of mind?

Imagine you could wire your brain for __________! (Again, fill in this space with the outcomes you desire!)

Imagine raising your fist to the sky and shouting, "Yes, I have made it!"

Again, *Get Wired for Success* is perfect for you.

So, let us look at what is inside
Get Wired for Success!

First, there are four chapters to this book and four matching units in the online course, each integrated and sequential, with each building on the one that came before it. The order is important; it is like building a skyscraper—you need strong foundations if you are going to reach for the sky.

In Chapter 1: How to Wire Your Brain for Success

- Learn how I crushed my mindsets and why that changed everything. (This was my big breakthrough).
- Discover how your brain works so that you can turbocharge it! This is neuroscience made easy—I explain the science to you in simple, easy-to-understand terms and language. There are no shortcuts here. Your knowledge of this science is essential. It will give you critical self-belief: "Ah-ha! I can do this!"
- Next, get The 12 Keys to Wire Your Brain for Success. This is the real nuts and bolts of the course and why it is so different from any other business or personal development program. I will clearly explain how each Key relates back to your business success and how the Keys will open doors to a new way of thinking and feeling, a new way to see your world, and a new life!

In Chapter 2: Dream Big! Then Turn Your Vision into Results

- Unleash your imagination and creativity! Discover how to create a crystal-clear vision of your dream business life and of you, living a life you *love*!
- Next, get my Seven Rules for Real Results, and that is what you want—to transform your dreams into reality. Without the vital Seventh Rule, you are doomed to mediocrity or failure. As we know, failure sucks—and so does mediocrity!
- Then, use my Four Magnifying Tools to add speed and power to your vision. Get to your dream destination as easily and quickly as possible.

In Chapter 3: On Fire with Purpose, Power, and Passion

- Use a Three-Point Plan to find your Life's True Purpose—the secret to unstoppable motivation.
- Bored or burned out? Get the Four Rules to Turn Work into Pleasure.
- Discover your Six Power Words—vital energy for courageous, decisive, and authentic actions.
- Learn how to Ignite Red-Hot Passion—inspire your investors, lead your team, and persuade your clients.

In Chapter 4: How to Master Your Thoughts and Emotions

- You have heard about the magnetic X factor that celebrities seem to radiate. Discover the X factor vital for business success, how to get it, and how to maximize its effect. Do not miss this—it is 50% of your results!
- People spend millions on gurus in the personal development space. This is what they *should* teach, why it works, and how to use it: The Three Stages of Personal Transformation.

- Are you blocked by your past? Discover the "software" to Program Your Mind for ________! (Insert the outcomes you desire!)
- Bonus! Conquer the pressure and overwhelming nature of busy and chaotic modern life with my Emotional Mastery Formula.

In the Conclusion

- I set you a challenge: to unleash your amazing potential.
- You will want to know if wiring your brain for success really does work. I will give you proof—The Three Signs of Success—which will tell you that you are on the right path. It is so exciting. Be inspired!

Does that sound great? It gets better!

Turbocharge Your Progress with *Get Wired for Success—The Course!*

The companion online course transforms the information in this book into *your* journey of discovery, knowledge, and insight. We all gain knowledge in different ways, whether by reading, looking, listening, or doing—and I have each mode of learning covered.

Each unit in the course has a set of superbly presented videos. Watch never-before-seen graphics of how your brain wires for the better (or worse). See carefully curated images that illustrate how each unit affects your business success (or lack thereof). There are transcripts of each lesson for those who want to read, as well as audio files for those who prefer to listen, perhaps making use of your time on the way to work.

Finally, and most importantly, there are worksheets so that you can learn by doing. Called The 21 Steps, these exercises will help make your journey, growth, plans, business strategies, and tactics a positive part of your life.

The 21 Steps are carefully designed to make you *think* about and understand *how* to create the business life of your dreams and live a life you love. By doing The 21 Steps, you will be *actively wiring your brain for success*. Importantly, if you do not do the work and complete The 21 Steps, you will not gain the maximum personal benefit and opportunities within the course.

Here is another bonus: You get lifetime access to The 21 Steps within the secure online platform. Your answers are saved to revisit, edit, and improve as you accumulate experience. In addition, each of the steps can be downloaded as an editable PDF document. To test your progress, there are quizzes at the end of each unit.

There are several themes within the course.

First, this journey of discovery is holistic. I am writing about the *whole* as well as the component parts—the whole of business and the whole of life, and how each part of your business and life relates to one another.

Leadership is not separate from marketing, and marketing is not separate from money, and so on, and they are not separate from your life. They are all part of a whole—you as a business owner, entrepreneur, leader, or manager and you as a person. So, at the start of each unit, I give you The Big Picture—a holistic view of how each part fits into this whole creative process.

Second, at the end of each unit, I give you a Keep It Simple summary to review the key facts and quickly revise or refresh when needed.

Third, I give you the key benefits realized by your community and by you when you master each unit. This *is* inspiring stuff!

The process is especially designed to maximize harmony in your business life; your interactions with your team, investors, clients, and suppliers; and your personal life. One part of this process is not at odds with or risking any other part—quite the reverse is true.

My results and the outcomes I hope you achieve—a business life full of purpose, passion, and success and a personal life of freedom, wealth, and happiness—are the result of many factors working in synergy. Each part is designed to augment the whole in a never-ending upward spiral of business and personal improvement and growth.

When everything is integrated and holistic, harmonious, and synergistic, wonderful things happen. That sounds *so* cool, does it not?

Does the journey to business success involve work? Of course, it does—a lot of it! We all know success—as measured by having the time and money to do what you want—does not come without work. However, when you get "it"—and I will explain "it" in the *Conclusion*—work becomes easy, joyful, and an integral part of your life. You begin to flow effortlessly from one moment to the next, each seamlessly linked and all part of the journey that is your life.

Now, is *Get Wired for Success—The Course* a good fit for you?

Would you like your business life to feel better, lighter, smoother, faster, deeper, and richer?

Would you like to be the star of the movie that is your life?

Would you like to transform your thinking and emotions and the way you see your world at the "flick of a switch?" Would that be *awesome*?

Then *Get Wired for Success—The Course* is a great fit for you. It will teach you these very skills.

Do you need to involve others in your decision here? Do your business partners or life partner need to do the course? It would be great if they do! Alignment with your partners is of so much benefit.

Do you have time to do the course? To make sure you are not overwhelmed, one unit is delivered each week. Each unit has four to five videos that are 10–30 minutes in duration, so they can be easily fitted into a busy schedule. Together with The 21 Steps, you should allow three to five hours each week for course completion. Can you invest this time in your future and success?

Does the course have to be completed in four weeks? No. You can do it in your own time and at your leisure.

Is there any better time to do the course? No. It is appropriate irrespective of your level of business experience or the size of your business.

Do you need to own a business? No. It is perfect for entrepreneurs, managers, leaders, CEOs, directors, consultants, coaches, accountants, dentists, software programmers, or anyone else involved in business. It is ideal for someone who wants to be more successful or is just starting out.

This is my guarantee:

1. Purchase *Get Wired for Success—The Course* and you will receive massive, amazing value.
2. Complete the online worksheets for Unit 1 and one other unit. You will look back and think, "Wow, what a sensational, mind-opening, and life-changing difference."
3. Do all four units of the course and The 21 Steps online, and you will be *actively wiring your brain for greater success in business and life*—a bold and amazing guarantee!

If you fail to receive these outcomes, e-mail me within 60 days of purchase at support@getwiredforsuccess.com to claim your 100% money-back guarantee. What value do you place on this knowledge, backed by such a rock-solid guarantee?

Be sure to check out the thank you page at the end of this book and follow the link to receive a big discount on the purchase price of *Get Wired for Success—The Course*!

Now, it is time for Chapter 1. Enjoy!

CHAPTER 1:

How to Wire Your Brain for Success

To create the business life of your dreams!

"The key to success is to focus our conscious mind on things we desire [and] not things we fear."
—BRIAN TRACY, American television host.

"Man cannot discover new oceans unless he has the courage to lose sight of the shore."
—ANDRÉ GIDE, French writer, humanist, and moralist; 1947 Nobel Prize winner for literature.

IN CHAPTER ONE

- The big breakthrough: How I crushed my mindsets and why this changed everything.
- Neuroscience-made-easy: How your brain works so that you can turbocharge it!
- The 12 Keys to Wire Your Brain for Success: This is the real nuts and bolts of the book and what makes it so different from any other business or personal development program.

THE BIG PICTURE

Let us project into the future and make one safe assumption: You will change. You are now different from the person you were as a baby, a child, or an adolescent. If you are an older adult, you are different to who you were as a younger adult.

Naturally, you will change physically throughout the years, but here is the thing: Your thoughts and emotions will change, too. It is not just a question of aging. Your brain changes *every second* in response to your environment.

The world is also changing, at an ever-increasing rate. The rate of change has never been as great as it is now, and today's pace will be the new slow in the near future. Rapid change is your reality. The critical question is this: *How* will you change? Will you be buffeted around by forces outside your control or will you drive your own destiny?

As you progress through your life, will you achieve your hopes and dreams? Do you have the personal power to make your dreams come true?

As a businessperson, can you create a business life full of purpose, passion, and success? If you are a business owner or entrepreneur, do you have all the skills to create an inspiring, vibrant, exciting, funny, compassionate, engaging, intensely interesting, progressive, and highly profitable business? With such a business, think of the amazing difference you would make to other people's lives. Think of the impact that would have!

Can you create an amazing life, with the time, wealth, and freedom to do what *you* want to do, and, yes, with real happiness, too? Imagine realizing and living your extraordinary potential; imagine this being your future—that would be wonderful, would it not?

Here is another critical question: In your heart and soul, do you genuinely believe it is possible? You will be familiar with the fact that we can learn to change our *physical* body and, with it, our performance. For example, an unfit person decides to do something about their fitness and joins a gym. With some coaching and hard work, they look different—perhaps slimmer and more muscular—and they can sustain higher levels of exercise or lift heavier weights or stretch and bend more flexibly. They have *learned* to change their body shape and performance, and they have likely improved their health, too.

Yet, *learning* to change the very way you think, your emotions, and your view of the world by wiring your brain for success? Other people have created such a life *by the way they think, the way they feel—with their emotions—and the way they see their world.* I have *taught* myself how to think, feel, and see that way, but what about you?

Before I wired my brain for success, I had to overcome a major obstacle, and you may also face it, so let us tackle it right now: Our mindsets, self-limiting beliefs, anxieties, and fears are part of our reality. We cannot imagine life without them; they seem to control or dominate us, and we cannot seem to escape them.

In the *Introduction*, I shared some of the negative mindsets going on in my head at the time, and the beliefs, anxieties, and fears that drove my behavior—my fear of financial failure, of conflict with my team or clients, of insecurity about the future—all blocking success. However, *positive* mindsets can also block progress!

Even when I was close to enjoying real success, I felt as though I was chasing a dream, and dreams do not come true—another false belief. Self-sabotage! My team seemed happy. We became a top-performing business. It felt like it could not get any happier or better—more false beliefs, this time generating complacency instead of growth.

Mindsets, both negative and positive, are a curse. I used to believe that my mindsets were just a part of me—like my eye color or the shape of my nose—as much a part of my future as they have been a dominant part of the past. They were my "baggage," carried around for the rest of my life, blocking all the dreams I had.

At any time, I can recall those mindsets and memories of my past—in some cases, painfully so. The baggage has stayed with me.

So, ask yourself what "baggage" *you* carry. Like me, you *cannot* leave your past behind. The baggage of negative life experiences remains with you in memories as you read these words. In part, through the life lessons they have taught you, they have made you the wonderful person you are.

How can you *learn* to leave this baggage behind? How can you *learn* to change this default setting, this imperfect status quo, this balance between positive and negative thoughts? Indeed, how can *you* crush *your* mindsets?

I have shared some of my early negative mindsets with you. Now, I invite you to identify your own.

What are *your* negative mindsets—your "baggage?" How do they limit your business and personal life and your growth? Do you confuse them with reality? What is holding you back from realizing your amazing potential, from real success, and from what you deeply want to achieve in your life?

Take the first step! Begin to Wire Your Brain for Success with Step 1 in the online course!

In Step 1, record the negative mindsets, self-limiting beliefs, insecurities, and even anxieties and fears that are blocking your amazing potential. We will return to Step 1 in Unit 4 of the course (How to Master Your Thoughts and Emotions).

Now, this was my "big breakthrough": How I crushed my mindsets and why that changed everything.

Picture this scene: I was holidaying with my wife, visiting her brother and sister-in-law in England. My brother-in-law had been unwell, so we all decided to rent a cottage and have a well-earned break in the beautiful English countryside.

Back home was the reality of my business, lurching from one crisis to the next; back home was over $1.1 million of debt; and back home was the awful clinical anxiety and insomnia that was threatening to overwhelm me and destroy my health and well-being. This holiday was not only to support my brother-in-law but also an escape from my awful reality

One day, we were all sitting in the living room of the cottage, reading books. I had pulled a book out from my suitcase that had been a birthday present from my son. It turned out to be a life-changing gift. The book was *Quiet Leadership* by David Rock,[5] a business management book to help business managers improve their own and other people's thinking—perhaps not everyone's choice for holiday reading!

I read Chapter 1, just 27 pages.

Suddenly, I became aware that my heart was beating faster, but not from stress or anxiety; this time, it was from excitement!

I looked up from the book and stared out the window, thinking, imagining, and playing with possibilities.

Chapter 1 of Rock's book looks deeply into how our brains work, summarizing cutting-edge discoveries and works by many leading scientists and psychologists. Based on proven science, there were two big takeaway messages for me:

1. Instead of being set, the internal "wiring" within your brain that controls your thoughts and emotions constantly changes.
2. There are strategies and tactics you can use to take control of this change.

This was my eureka moment—my breakthrough! Two thoughts crashed into my consciousness, one after the other, like two blinding flashes of light:

1. If my brain was *not* set, maybe I could cure my anxiety and insomnia.

---

5 David Rock, *Quiet Leadership* (New York: HarperCollins, 2006).

2. If my brain was *not* set, maybe I could turn from a being a mediocre businessperson into a successful one—maybe even highly successful.

I looked at my companions. I felt like shouting, "Hey! This is important! I can change my life! This is astounding!"

"Hey, maybe I can go from anxiety to calmness, from low self-esteem to confidence, and from weakness in the face of conflict to courage!"

"Hey, maybe I can go from $1 million in debt to owning a multimillion-dollar business!"

My brain was screaming messages at me. I could leave my baggage behind! I could crush my mindsets.

My companions continued reading their books, unaware of the thoughts cascading through my brain, altering, molding, and twisting it into places it had never been before.

Yes, here was the evidence right inside my head, at that very moment in a cottage in the south of England: My brain was changing!

This knowledge changed *everything*. Yes, design my future! Yes, drive my destiny!

This is what I learned and what you need to know—how your brain works so that you can turbocharge it!

Your brain has over *100 billion* brain cells. Each brain cell searches out for and connects with other brain cells. At any point in time, each cell can have up to *100,000 connections* with other brain cells. Disconnections occur too, and the whole process of searching for, connecting with, and disconnecting from other cells makes our brains work.

At any point in time, over *one million* new connections and disconnections are happening *every second*, with over *300 trillion possibilities* for connection![6]

Just like in a supercomputer, little currents of electricity run along the connections between brain cells, and that makes our brains work, as we think and plan, feel our emotions, and interact with our environment using our senses. Pretty amazing!

Your brain is constantly changing and adapting to your environment because your brain cells are "searching, connecting, and disconnecting" all the time. As you read this paragraph, your brain will have made millions and millions of new connections *and* disconnections while you remain blissfully unaware of the process.

With such extraordinary complexity, no two brains are alike. Even identical twins' brains become quite different as they respond to their different environments.

So, here is a test: How many new brain cell connections happen every second?

If you immediately knew the answer because you have just read it, or if you had to go back four paragraphs to find the right response, a very short-term memory had formed in your brain, immediately giving you the answer, or at least how to find it.

This memory occurs because brain cells form connections with other brain cells that become *enduring*. The individual brain cells that connected to make up the memory of the answer to my test created a *network* of cells that endured until I set you the test.

---

6 Adapted from Rock, *Quiet Leadership*.

Our life depends on brain cell networks and their enduring connections.

Let us look at some other networks. For example, how you bring a cup to your mouth to drink is a brain cell network. If the drink is too hot and you pull the cup away when it touches your lips, *that* is also a network. Your memory of your childhood home is another one. Remember your bedroom in that home? That is a network, too. The route to work is a network, and so is knowing the words to your favorite song. If I say its title, you remember the tune. That is a network.

Get the idea? Even an idea is a network!

Some networks are short-lived, such as what someone said to you last week. Others are long-lived, such as those about your childhood home. As disconnections are also happening continuously, some networks are lost—for example, a phone number you used to remember but have since forgotten. *Disconnections make space for new connections!* They are just as vital as connections.

Some networks are important, with extraordinarily strong connections between the relevant brain cells, and they remain within our brain, helping us to navigate and survive in our complex world. For example, the brain cell network telling you to pull the hot cup away from your mouth is important. When you are driving a car, applying the brake in an emergency is important, as is your ability to read. The brain cell connections forming these networks are not only long-lasting but strong! These brain cell networks are "hardwired" in our brain, like memory chips in a supercomputer.

Of course, at any one time, we are unaware of our networks, but they govern our lives—when we wake, when we work, at play, at mealtimes, when we go to sleep, even when we dream. Billions of brain cells are busy searching, connecting, and disconnecting with billions of other brain cells, all the time making the networks that drive our lives.

Here is the critical point: *Our brain cell networks are not set.* They are changing all the time, as our brain cells search, connect, and disconnect. They are infinitely variable. They are *plastic* (in the sense of "molding or changing," not to be confused with the material with which we are so familiar). The scientific study of this plastic nature of our brains is the relatively new and little-known science of *neuroplasticity—the science of wiring and rewiring your brain.*

Following my holiday in the south of England, I returned home with one positive emotion dominating my thoughts: faint, tentative, trembling *hope* for a better balance between my negative and positive thoughts and emotions, for improved health and well-being, for business success, and for a better life!

I realized that many of my mindsets and self-limiting beliefs were networks that were *hardwired* in my brain, and I could not seem to escape my awful reality.

Rock's book was the spark. Gripped by a sudden desire to know more about neuroplasticity, I was now fired up with curiosity!

Back home, I read book after book. Some were absolute garbage, penned by charlatans posing as experts. Others were pure gold, written by world-leading authorities in their field, generous giants sharing their wisdom for our betterment. I was lucky that my training as a scientist enabled me to understand the science and sort the gold from the trash.

I soon found that I needed to branch into other subjects—positive psychology, visualization, motivation, emotional intelligence, meditation—for a more complete understanding of what I was trying to do.

What *was* I trying to do?

Back then, I would not have been able to verbalize an answer. Only with the wisdom of hindsight do I now know. I was discovering how to totally transform myself, to crush my anxieties and fears, to unleash my potential, and to create the business of my dreams so that I could live a life I *loved* and powerfully contribute to my community.

Applying this extraordinary science, this new knowledge, and these wonderful insights to both my personal and business lives totally changed me in the most positive and energizing ways. I crushed my mindsets. Slowly, my dreams became reality—transformed! Wow!

The catalyst for this change to my thinking and emotions—from being externally controlled and reactive to being internally designed and directed—was this: I learned to wire my brain for success. I learned to direct my own neuroplastic change! This skill is now called *self-directed neuroplasticity*.[7]

Now, it is *your* turn!

Not only can we decide to improve our physical performance by going to the gym, but advances in neuroscience now tell us that we can also turbocharge our brain—a gymnasium of the mind.

7 When I write about *in*creasing the number or strength of brain cell connections associated with positive thought patterns or emotions, or about creating new positive networks, I use the term *wiring* our brain. When I write about *de*creasing the number or strength of brain cell connections associated with negative thought patterns or emotions, I use the term *rewiring*.

From my years of study, I have collated the work of world-leading neuroscientists and psychologists and simplified and summarized it for you here. It is the best of the best—The 12 Keys to Wire Your Brain for Success!

This information is exciting—it has the capacity to change your life for the better, forever.

Are you ready to make that change? Are you ready to alter your default settings? Do you want to intentionally change the imperfect balance between positive and negative thoughts that drive your life, seemingly forcing you down reactive, undirected, and uncontrolled paths? Do you want to crush your mindsets? Yes? Then, wire your brain to get the results you want. This is how you do it.

THE 12 KEYS TO WIRE YOUR BRAIN FOR SUCCESS

The First Key:[8] Hardwiring is difficult to undo, but it can be undone

Many brain cells are involved in a hardwired network, and the connections between them are strong. Deeply embedded or important brain cell networks tend to endure.

At first glance, the First Key *seems* to be a barrier to success, but remember that hardwiring is there for a reason. Our lives depend on many hardwired networks; we simply cannot do without them.

However, other hardwired networks really are a barrier to success—notably our weaknesses and the negative mindsets and beliefs and the baggage we carry from our past.

8 Keys 1–3 were adapted from Rock, *Quiet Leadership*.

Each one of the negative or inhibiting thought patterns, self-imposed limitations, and anxieties and fears from my past was a different brain cell network. Thoughts of failure and anxiety and my physical responses to them are all deeply held within my brain, and some are extremely hardwired. As I write these words, they are still clear as a memory.

Here is the key point: If over one million new brain cell connections are happening every second, about *one million disconnections are also happening!* If hardwiring is difficult to undo but *can be undone* with the science of neuroplasticity, then the brain cell networks of your negative thoughts, beliefs, and mindsets—even of your anxieties and fears—can be undone. If your self-beliefs can be *undone and reassembled in a vastly better light*, what are you capable of doing?

Be suddenly unlocked from your past. Fully engage with the present. Imagine a bold, new future! Be liberated! Unleash your amazing potential! Feel the ground shift!

The *good* news is that even though the hardwiring of strong and enduring negative thought patterns, beliefs, and emotions may be difficult to undo, by using the other 11 Keys—all based on the science of neuroplasticity—rewiring may be very quick and occasionally lightning fast!

This First Key explains why improvement in any endeavor can be slow—so slow, it can be imperceptible—giving rise to the whole concept of mindsets. We are all familiar with phrases like "her mind is set" or "he has got the wrong mindset!" Here are four reasons why mindsets—even positive ones—can block success.

First, our mindsets place our own preconceived notions on what is or is not possible, just as I did years ago. How often have you heard others say, "I could never do that," when the challenge before them seems so easy to you?

Who makes up these rules that limit our ability? While our relatives and friends, teachers, fellow workers and employers, community, and the media all influence us, the fact is that we do! *We* make up our own rules, often unconsciously so. Even using a phrase such as "all you need is a winning mindset" places an upper limit on progress. In other words, once you have a winning mindset, you cannot do any better. *Who* said you cannot do any better?

Second, have you ever woken up with the most amazing mindset? "Wow, today I feel great! I can conquer the world!" Then, one hour into your working day, that incredible mindset is in tatters or forgotten, as all the troubles of the world (and of your business) land on your shoulders. Two of your team members have just had a fight, your best client is raging on the phone, upset with your service, and last month's sales figures are down. Now, where was that positive mindset and what good will it do you in that moment?

Third, much is made of mindset training and having a growth mindset. I am all for growth, but it is important not to be fixated on it. Especially if market conditions are adverse, it may be more appropriate in business to maintain rather than invest in growth, to reflect and plan and not charge ahead.

A growth mindset is synonymous with a commitment to personal change, but if change is not in the right direction or for the right reasons, or if one lacks the power to effect it, then growth for the sake of growth can be exhausting or demotivating. Worse, you could burn out, disillusioned with lacking progress or advancement. You may also be so set on your path to business or personal growth, other important aspects of your life are neglected.

Finally, the word 'mindset' itself fails to address the fact that although progress toward a better set of thought patterns, beliefs, and emotions may be slow or difficult, the science of neuroplasticity proves that it *is* possible—at a million connections *each second*, with over *300 trillion* possibilities for connection. The very word mindset is a misnomer! *The mind is not set!*

It is time to get rid of tightly held mindsets, including positive ones. We need a model based on the *science* of how our brains work, not based on an abstract concept such as mindsets. We need a model that makes us *think*, one that is more dynamic, more flexible, and indeed more *plastic* than mindsets!

But wait a minute! We have just learned that your brain *is* plastic. Voilà!

There is also a wonderful upside to the First Key. All your strengths and positive thought patterns are already hardwired, preprogrammed for success. Just like upgrading software in a computer, by using the other 11 Keys, you can increase the number, strength, and durability of brain cell connections responsible for your strengths, transforming them into *superpowers*!

The Second Key: Negative brain cell networks become even stronger if we continue to focus on the problem. Likewise, positive networks become even stronger if we focus on positive outcomes and opportunities

In your mind, what you focus on grows bigger and bigger. This is because more brain cell connections are recruited to make even stronger and more enduring networks. The negatives can be so dominating and so stressful that they totally dominate your thinking and emotions and (as I found) can even cause ill-health.

We have an instinct to search for answers to annoying or disabling habits. We all want to be better and solve our problems, but this search and focus often makes things worse.

So true! Back in my business, if I lost a client, I was not a good enough businessperson and had to try even harder not to lose clients. I was focused on loss and negative self-esteem. If I had to confront a team member, the more I tried to relax, the more anxious I became. I was focused on my stress and not the fact that the issue at hand was a golden chance to train the whole team. When I received a disappointing set of financial figures, my focus was on the bad news and not the opportunity that lay within the figures. My body language was extremely negative, and this influenced my team—morale dropped.

We can be hardwired for disappointment.

Some of you will go to business management conferences or read business management books, only to return to your business and find that nothing changes or, even worse, deteriorates. This is because you have not wired in new, strong, enduring networks to support your new knowledge, or you still focus on the negatives.

There is a wonderful flip side to the Second Key: Positive brain cell networks become stronger if we focus on positive outcomes such as opportunities, goals, rewards, achievements, positive relationships, and emotions.

This flip side explains why some self-help gurus are so successful—they teach their adherents to focus on the positives, and more power to them!

However, it does help to know the reasons *why* you are successful. Without knowing the reasons *why*, it is easy for those negative networks to remain, and without knowing *what* to do about them, they can once again dominate our thoughts and actions, returning us to square one. Then, we lose faith in the teachings of the self-help guru of the week, moving onto the next one.

Despair not! Fortunately, there is a Third Key.

The Third Key: It is easy to create new wiring

Your mind is not set! It is continuously rewiring at 1,000,000 connections every second. New brain cell networks are constantly being made. The trick is to *intentionally* design and create new networks to our massive advantage to *direct* positive neuroplastic change.

As I wrote earlier, I used to believe that my thought patterns could not be changed. Stuck with who I was, I was mired in mediocrity. How wrong I was! My brain was suddenly alive with possibilities—I could reach for my dreams!

It is the *intentional creation* of a positive future that allows us to escape from the baggage of our past. For me, the realization that I could leave the past behind was exciting, inspiring, motivating, and, yes, liberating.

Thinking more deeply, I realized that through my actions (or inactions), I could also create positive (or negative) brain cell networks *in the brains of other people*, specifically (and importantly) in the minds of my team, investors, suppliers, clients, and even my family and friends. Their brains are plastic, too!

Now that was a mind-blowing realization. Think of the implications for success (or failure) of a business—*your* business—and of a life—*your* life!

It was a light-bulb moment!

So, what happens to the unwanted and disabling negative brain cell networks of our past? Remember, billions of brain cells are constantly searching for other brain cells, connecting and disconnecting. Again, the disconnection process is critical. Unwanted networks become less connected, less enduring, weaker, and, ultimately, a thing of the past. They become just a benign memory, no longer controlling us. We will return to the process of disconnection in the Tenth Key.

The past is important. The lessons learned from it are essential to our future success and even to our survival. We do not want to repeat our mistakes. Reflect on the past, but do not dwell on it. You cannot change your genetics or your past. It is over and gone. All previous environments and experiences are now history.

Learn from the past, but let the past "be." Instead, design your future; create the business of your dreams; and live a life you love!

The Fourth Key:[9] Both the strength and number of connections between brain cells of a network increase with repetition, irrespective of whether the network is a positive or negative one

This is the scientific reason for why we improve over time with practice. Think about elite performers. A concert pianist or a top athlete does not become an instant success. They practice over, and over, and over, for many years, before they become a master of their art or endeavor.

---

9 Keys 4–7, 9, and 10 were adapted from Michael Merzenich, *Soft-Wired: How the New Science of Brain Plasticity Can Change Your Life* (San Francisco: Parnassus, 2013).

There is a saying in the field of neuroplasticity: "Brain cells that fire together, wire together."[10] Within the brain of an elite performer, the brain cell connections critical for their amazing performance become greater in number, and the connections between them are much stronger as they practice, practice, and practice some more—until mastery is achieved. They are hardwired for success!

Those elite performers "play to their strengths," but do you? Do you focus on all the positive thought patterns that make you the unique and amazing individual you are? Do you not only focus on them but also *practice* your strengths, or do you just take them for granted?

Remember, it is the combination of your positive beliefs and attitudes (along with your negative ones) that has shaped your subsequent behavior and actions, which, in turn, have got you to where you are right now. Play to your strengths!

Just as we can intentionally design and create *new* positive brain cell networks to replace negative ones, we can intentionally strengthen our *existing* positive networks with time and practice to make them even better. Again, turn your strengths into superpowers!

Mastering skills in any endeavor—and specifically in business and entrepreneurism—takes time and practice. For me—from the low point of undirected, reactive chaos—I did not acquire the skills it took to create my dream business overnight; it took time, energy, practice, and patience.[11]

---

10 This is a popularized summary of Hebb's theory in neuroscience. It is not technically true, as brain cells fire one after another rather than together, but it does paint a vivid picture of the process.

11 We will look closely at how to make your existing positive brain cell connections even stronger and more enduring in *Chapter 3—On Fire with Purpose, Power, and Passion*—and *Chapter 4—How to Master Your Thoughts and Emotions.*

We think nothing of practicing repeatedly to master a hobby, an artistic endeavor, or a new language. We take for granted the intensive training required for elite sporting teams to reach the pinnacle of success, yet how often do we *practice* business skills or *coach* our teams to deliver on the promises we make to our clients? We practice in the arts and we train for sport—*now practice and train for success!*

To return to our negative thought patterns, this process of brain cells "wiring together and firing together" further explains why focusing on a negative aspect of our behavior or beliefs makes things worse. It is like practicing for failure! It also explains why criticizing people seldom improves them. Rather, criticism creates negative brain cell networks!

It also explains why it can be so difficult to implement new knowledge in our businesses—especially if you do not know how to change the way you think, let alone the way other people think, behave, and act.

Now, let us use the Fourth Key to review, as review is a form of repetition. These points may present an entirely new paradigm in the very way you think:

- Ignore your negative thought patterns and let them be.
- Design new positive thought patterns to replace the negative ones. Strengthen them with practice.
- Focus on existing positive thought patterns and strengthen them with practice.

We have a natural tendency to focus on our negative thought processes and take our positive thought patterns for granted. We limit what we believe we can do and downgrade our capacity to take on new skills. Think of the implications if we can flip this around. Think of the implications in areas such as adult learning and coaching or in managing and leading others. Think of the implications for business!

Now is the perfect time to take Step 2 in the online course!

In Step 2, identify the positive thought patterns and strengths that make you the amazing and unique person you are. Importantly, we will return to the Step 2 worksheet in Unit 4 of *Get Wired for Success—The Course*. Discover how to turn your strengths into superpowers!

The Fifth Key: Change must be desired. The greater the desire, the greater the possibility of wiring or rewiring

When you are fired up for change, when your brain is screaming in mental pain or intensely desiring pleasure, when motivation and engagement are high, the brain releases chemicals that facilitate the process of brain cell connection and disconnection. Wiring or rewiring is easy.

On the other hand, if motivation is *not* there—if you are disengaged, disinterested, fatigued, or exhausted—then wiring or rewiring is more difficult.

Remember that awful financial and emotional pit I described in the *Introduction*? Eventually, I dug myself out, and year after year, my business improved. It was as though this was the natural progression of business—that I had found the magic formula to success and that the good times would keep on rolling.

Yes, I had the Midas touch! The adrenalin was intoxicating.

Then, one month, without warning or identifiable reason, the figures started heading south, and the downturn continued each month. Value per client visit, number of client visits, gross earnings, profit—all the key indices were down and kept heading that way.

At first, my brain was frozen with inaction, perhaps even with shock. I tried to explain it away—seasonal variation, a downturn in the national economy, or a marketing campaign by my competition. All or none of those reasons may have been valid, but the fact remained: My business was beginning to hemorrhage money.

If this downturn continued, I was faced with losing *everything* after working *so* hard for it. For a week or two, every time I faced the reality of negative financial figures, I felt the old familiar shiver of fear pulling at the edges of my brain, trying to unhinge me, to reopen a door to a horrible past—to clinical anxiety, insomnia, and failure. However, I had my 12 Keys this time!

"No, no, NO! I will not slide down into that pit again," I told myself. "Right, where is the opportunity in this downturn?"

"What is done is done; learn from it, but let it be. Focus on the positives. Create new brain cell networks!"

Suddenly, my brain kicked into action. It was like jump-starting a car or applying a defibrillator to a heart in cardiac arrest. Without warning, and with renewed energy, new marketing ideas began to pour from my brain, too many to enact at once. I began to manage my team better, communicate with my clients better, and develop better performance metrics. Equally quickly, the financial figures began to climb back into the black!

On reflection, my brain had been screaming in mental pain. I have no doubt that the desire to avoid loss and pain is a stronger catalyst for wiring your brain than the desire to improve things when everything is going well.

My training, mental strength, and resilience prevented negative thought patterns and emotions from overwhelming me, and my personal strengths of creativity and communication were enhanced.

This is the intriguing point: My business was better *after* the downturn than before it, and I was a better leader and manager. The fierce *desire* within me to avoid loss created entirely new initiatives and skills, placing me in a stronger position. New brain cell networks for marketing, leadership, and analysis were wired into my brain, seemingly without effort.

Vastly improved control of my stress and anxiety had created the capacity for creativity—to return to the computer analogy, just like installing a new drive creates more space for processing.

Desire is a massive catalyst for accelerated neuroplasticity.[12]

Conversely, if someone is demotivated, disengaged, disinterested, fatigued, or exhausted, wiring or rewiring is more difficult. Lack of desire makes efforts to facilitate change less effective. You cannot easily change someone if they have no will to change.

---

12 In *Chapter 2—Dream Big! Then Turn Your Vision into Results*, you can boost your levels of desire with simple, easy-to-remember strategies. Use them every day in your business and personal life, hijacking your brain's ability to wire and rewire in response to desire. In *Chapter 3—On Fire with Purpose, Power, and Passion*, take a deep dive into the subject of motivation. You can create a compelling purpose for yourself so as to fully engage with the present. You can also clarify what is important to you so that you have the personal power and courage to motivate others.

The Sixth Key: Remembering goals, models, and strategies guides positive wiring

If you have goals, models, and strategies in your business or personal life, each one is a brain cell network.

This alone is an extraordinary fact. Imagine this: Abstract concepts such as the thought of a goal have actual physical substance in the form of thousands of connections between brain cells. Imagination is very real, even if it is only in your mind.

Referring to the Fifth Key, if your goals and models are fiercely desired, they are composed of strong and enduring brain cell connections.

We all know the difference in our "feelings" between being half-hearted about a goal and being totally focused on it. Now, you can visualize the difference in terms of the number and strength of connections involved in the brain cell network of that goal. Indeed, the greater the clarity or definition of those abstract concepts, the greater the number of brain cell connections involved.

Those elite performers I mentioned have specific *goals* guiding their practice until mastery is achieved—be it a gold medal in the Olympics, a team trophy, or the thought of holding an audience enthralled during a concert. They have precise *models* in their mind outlining or illustrating exactly how they intend to perform—their body position, their fine movements, and even control of their breathing and mental state.

In competitive sports, top athletes will have *strategies* in mind as well—what to do or not to do if the weather changes, if the other competitors are winning, or if they are not fully engaged. These strategies are all defined as networks within their brain. *Their* goals, models, and strategies define their elite performance.

Your goals, models, and strategies (or lack of clarity about or absence of them) define your performance in business and in life.

The Sixth Key explains my initial success with my business. As I outlined in the *Introduction*, I knew what I wanted: a model business. Slowly, without being aware of the neuroscience underpinning my desire, my dream became reality. The model in my mind—defined by strong and enduring brain cell networks—governed my actions that, in turn, created the results I wanted.

In Units 2–4 of *Get Wired for Success—The Course*, you can construct extremely specific, positive, effective, and motivating networks in your brain that will define your goals, models, and strategies. They will describe the what, where, how, and why of your future and build on the personal power needed to achieve your dreams. They will turbocharge your beliefs and emotions, the actions you take, and the results you get.

You can learn core strategies to help create positive beliefs and emotions in others, too. You, your team, and your clients can go from strength to strength. Wins and win-wins create bigger and better neuroplastic changes in an ever-upward spiral. It can be so exciting!

The Seventh Key: The greater the reward (or loss or failure), the bigger the wiring or rewiring

Every big win and failure (even if only perceived) flood the brain with chemicals that boost wiring, making new, strong brain cell connections for the better—or the worse!

We have all seen individuals go from strength to strength, and we put their good fortune down to luck or being "genetically gifted." Sadly, we have also witnessed people caught in a negative downward spiral and are dismayed at how they can get even the most basic things so wrong. The brains of both sets of individuals are undergoing massive neuroplastic change, and they do not even realize it. One is being wired for success by the incremental rewards of upward progress, and the other is being wired for failure by repeated losses.

Of course, rewards play a large role in business. We see business owners and entrepreneurs throw themselves into their business with high energy for big future goals, only to burn out in the short term, never achieving their dreams because they never rewarded themselves *along the way*.

We see bosses drive their teams hard all year and then think that an annual raise and a meager Christmas bonus is enough to motivate their employees—and they wonder why staff turnover is so high. A simple word of thanks or an act of compassion or kindness can be a greater reward (due to the positive emotions created) than bonus systems or raises. A word of thanks or an act of compassion costs nothing!

We also see customers being rewarded for their purchases or brand loyalty with all manner of rewards, likewise unaware of the neuroplastic changes going on inside their minds.

What about you? Do you reward yourself appropriately for your hard work, or do you just keep working? Do you link your rewards to upward progress or are your rewards ill-conceived whims that jeopardize your financial position?

In *Chapter 2—Dream Big! Then Turn Your Vision into Results*, we will look at the role of both frequent, healthy, inexpensive rewards and occasional big rewards to drive positive neuroplastic change in YOU.

The Seventh Key gives you some insight into why gambling and addictions can be so difficult to control or cure. The occasional win for a gambler or the high of drug use for an addict are *big* rewards, strengthening the brain cell networks that powerfully control their addictions.

The Eighth Key:[13] Mental rehearsal, meditation, and mindfulness cause the same wiring in your brain as real-life experiences

This is a *stunning* statement.

Wiring for success occurs in mental rehearsal! New brain cell connections—including all those required for business success—occur even when you just repeatedly *think* about a desired belief, behavior, action, or emotion.

Have you ever watched aerial skiers rehearse their extraordinary moves just prior to a jump? They cannot make all the moves before the jump because their feet are still anchored to the snow, but they are making the jump in their mind. Then, they commit—their lives depend on it! They hardwire their brains to make the right moves, at the right time, and at the right speed.

You can do the same in your business world!

Importantly, rewiring also occurs when you meditate. I had used meditation to control bouts of disabling anxiety, so I knew it could be used to treat negative thought patterns and emotions. *But to create positive new brain cell networks* or turn my strengths into superpowers? Wow!

So, I designed and practiced meditations to wire my brain for success. I call this "applied meditation." Although we can search

13 The Eighth Key was adapted from Richard J. Davidson and Sharon Begley, *The Emotional Life of Your Brain* (London: Hodder and Stoughton, 2012).

for inner peace with mediation, we can also use meditation to find external success.

Mindfulness is the art of being in the present moment rather than worrying about the past or future. It is the skill of focus and precision rather than our attention being scattered in futile multitasking and distraction. It is the joy of taking one step after another in a perfectly designed journey rather than going around in head-spinning, tight, little circles, getting nowhere.

With the focus of mindfulness, thought patterns and emotions repeat in our minds rather than scatter, in turn creating stronger, more numerous, and enduring brain cell connections aligned with the outcomes you desire!

For me, the knowledge that you could wire and rewire your brain simply by mental rehearsal, meditation, or mindfulness was the single most transformational realization in my journey from undirected chaos to thoughtfully planned design, from anxiety to peace, and from ignorance to success. I used to feel totally overwhelmed in my business. I applied the Eighth Key to control that paralyzing, distressing feeling and regained calmness—despite the *insane* amount of business I had to deal with.

Discover how to create the business life of your dreams and live a life you *love* by using mental rehearsal, meditation, and mindfulness in *Chapter 4—How to Master Your Thoughts and Emotions.*

The Ninth Key: It is as easy to cause negative neuroplastic change in your brain as it is to create positive changes

Wiring and rewiring are facts of life. They are happening in your brain all the time whether you like it or not and whether you are in control of the process or not. You are unconsciously rewiring, but you can also consciously wire new brain cell connections, just as surely as you can install new capacity and speed in a desktop computer.

However, it would be too easy if all we had to do was "install" a "think-positive circuit board" or a "positive mindset chip" into our brains, and all our dreams would come true. Unfortunately, this is not the case. With all this talk in the self-help industry about positivity and positive mindset, how do negative changes occur?

Let us use a business example. Imagine you are at work. You are very tired and irritable, and you want to be left in peace. An employee asks you an annoying and trivial question. You snap back the obvious answer. Your employee walks away, with a look that says: "I will *not* ask that stupid question again."

As you were rewarded by getting the result you want—peace and quiet—the brain cell network or thought pattern that generated your response is strengthened by neuroplastic change, hardwiring your behavior, and making it more likely you will make the same response again.

However, because of your response and the negative feelings it generated *in your employee's brain*, they are less likely to bother you again, withholding potentially important information or great ideas. Your impatient answer is one small step in thousands of interpersonal interactions that happen in your business life each day, and each interaction makes up part of your business culture.

Culture is another core component of business success. Even if it is only small, that one strengthened, negative neuroplastic change can weaken the positive beliefs, behaviors, and actions that make up your business's culture.

Now, sum together all the positive and negative interpersonal interactions that occur within your business life (or in any business) daily, each driven by often-subconscious thought patterns or networks in your brain, and each strengthening or weakening with time. Are your brain cell networks, *and those of your team*, the result of thoughtful design or uncontrolled, reactive chaos?

If it is as easy to create negative neuroplastic change as it is to create positive change, can you see why it is imperative to have clear, positive, and motivating direction, strategies, goals, and models to follow? Without them, our brains can flip-flop between negative and positive change without making any progress. No wonder people become bored with their jobs! No wonder they become sidetracked by all manner of distractions! No wonder they become burned out! Think of the business and social implications.

The Tenth Key: With every positive neuroplastic change, the brain takes the opportunity to weaken connections in negative brain cell networks

Over time, as some connections between brain cells "that fire together" increase in strength and number, connections that "do not fire together" become weaker, culled from the brain cell network or thought pattern in question. As I eluded to earlier, the disconnection process is also critically important. The unwanted, negative networks in our thinking and emotions can become less connected, less enduring, and, ultimately, a thing of the past. They become just a benign memory, no longer controlling us.

This explains why the positive psychology movement, the billion-dollar personal improvement industry, and concepts such as a "growth mindset" have gained so much traction. In *Soft-Wired,*[14] Dr. Michael Merzenich states that

14 Merzenich, *Soft-Wired,* 57.

> "every moment of learning provides a moment of opportunity for the brain to stabilize—and reduce the disruptive power of—potentially interfering backgrounds or 'noise.'"

How could this play out in your business world?

To return to the example of your impatient answer to the annoying question from an employee, if your reply had been more considered—for example, patiently encouraging your employee to use their own initiative to find the answer they wanted—this would not only strengthen positive neuroplastic change but also weaken *the brain cell networks controlling your impatience.*

It gets better! Positive networks in your employees' brain associated with you and your business *will also be strengthened,* building their engagement, motivation, and loyalty. Any negative networks about you or your business *will also weaken.* In turn, their opinion of you will shift slightly, as their opinion is a brain cell network.

While I have used the example of an employee, the Tenth Key is equally applicable to your investors, suppliers, and clients, as well as to your family and friends. Wow! What an opportunity this extraordinary capacity of our brain for wiring and rewiring gives us!

The Eleventh Key:[15] Positive or negative emotions instantly increase the strength and number of connections in a new or existing brain cell network

Think of wonderful or happy occasions throughout your life: your first kiss, a wedding, the birth of a child, graduating from

15 Keys 11 and 12 were adapted from Loretta Graziano Breuning, *Habits of a Happy Brain* (Avon, MA: Adams Media, 2016).

school or college, a sporting victory, or a special holiday. These memories are etched into your brain as if the events were only yesterday. This is because the brain cell networks associated with the event instantly formed thousands of strong and enduring connections in your brain due to the *emotions* associated with them.

The same process happens with positive experiences in your workplace: Think of landing your dream job, having a big win, meeting kind or generous people, or having a great boss. The same process is in play, with emotions likewise etching strong and long-lasting brain cell networks of these events or people in your brain.

Now, think of sad or unhappy occasions in your life, perhaps an illness, the death of a loved one, or watching a disaster unfold. The negative *emotions* generated by these events will also create strong and enduring networks in the form of clear memories in your brain as the events unfolded. The same is true in the workplace: being fired, having a big loss, being bullied, harassed, or abused, or experiencing a poor boss, nightmare clients, or customers—each has left indelible memories in your mind.

In each circumstance, positive or negative, your brain has wired itself automatically, instantly, and profoundly *because of chemicals released in your brain when deep emotions are involved.*

You can utilize the positive aspects of the Eleventh Key to *massive* advantage!

Imagine all the interactions you have with your customers, clients, suppliers, and team members. What memories will you create in their minds, and what emotions will be associated with those memories—positive or negative? How will those emotions affect their subsequent beliefs, attitudes, behaviors, and actions toward you and in your business?

You wire other people's brains! You cannot help but do so. Once again, the implications are profound.[16]

The Twelfth Key: You can wire your brain to feel positive emotions such as joy, celebration, gratitude, a sense of belonging and, yes, happiness by creating new brain cell networks that release a cocktail of feel-good chemicals

The human brain has evolved over millions of years to enable us to successfully navigate our challenging and even life-threatening environments. Through millennia, emotions have been essential in this battle for survival.

Your ancestors needed positive emotions such as joy, happiness, and love to reinforce experiences that favored survival of the species—finding food and shelter, courtship, sex, and child-rearing. Their brains evolved to release feel-good chemicals that saturated pleasure centers in the brain, generating the positive emotions *and responses* they needed to survive.

Likewise, they also needed emotions of fear and unhappiness to avoid life-threatening situations and to learn from their experiences to survive. Their brains evolved to release other chemicals at times of danger or stress, saturating areas of their brains responsible for self-protection and survival, generating the negative emotions *and responses* they needed to survive.

We need both positive and negative emotions to navigate our world, even though it is a vastly different and more complex one

16 We use the Eleventh Key extensively throughout the companion online course as you map *your* path to freedom, wealth, and happiness.

than that of our primitive ancestors. It is balance that is important, and here is the critical point: By tipping the scales strongly in favor of positive emotions and interpersonal relationships, you will accelerate progress on your path to success.

By reducing the chemicals coursing through your bloodstream associated with chronic stress, anxiety, or fear, your health and well-being can also improve.

It is not just about happiness for you—you can generate positive emotions in others, too. You can benefit your team members, clients, community, friends, and family. It is a win-win-win situation!

Throughout *Get Wired for Success*, we will look at ways of creating positive experiences for you and others. Release feel-good chemicals for the good of all. Enjoy!

One important aspect of these feel-good chemicals is that they are quite short-lived in your brain. Their effect is often felt for only a minute or so. We are not built to feel happy all the time—if we were, we would miss the threats and dangers to our health, well-being, and survival. Learning to utilize feel-good chemicals in our workplaces is an ongoing challenge.

The Twelfth Key begs important questions: How *do* you tip the balance in favor of positive emotions? How do you transform negative emotions into positive ones?

How do you create a feeling of optimism when you are feeling pessimistic or joy in an environment of despair? How do you feel gratitude when you are feeling ungrateful about your current situation, or create feelings of abundance when there is scarcity all around you? How do you feel happiness when you are sad, calm when you are anxious, and love when you are fearful or full of hate?

Important questions indeed, for all these emotions can all play out in your business *and* personal lives!

We will discover how to experience wonderful and positive emotions when you need them most in *Chapter 4—How to Master Your Thoughts and Emotions.*

The 12 Keys represent a dynamic, flexible, and indeed plastic model for business and personal growth. It is important to note that while this model is based on the science of how your brain works, many other factors influence its performance. For example, your age, gender, and environment influence it, and the quality of your sleep, nutrition, exercise, and social support all profoundly affect the outcomes.[17]

The 12 Keys *are* the secret source of my success. With them, the pieces of the jigsaw puzzle I could not solve began to fall into place.

I could take control of my business and my life. It was in my hands—to succeed beyond my wildest dreams, wallow in mediocrity, or fail. Which road did I choose?

I learned how to design and direct my own thinking, emotions, beliefs, and behaviors and then integrated them into my life. I learned to become the very person I wanted to be. It allowed me to be the architect of my dream business and the designer of an amazing life.

Through this change, I have a business life full of purpose, passion, and success; I have the freedom to do what I want to do; and I enjoy real happiness.

17 If you wish to further optimize the performance of your brain, I highly recommend Dr. Sarah McKay's online course *The Neuroscience Academy.*

As part of my change, I am now much more considerate, generous, and thankful to others. I am a better manager and leader; I am calmer, kinder, gentler, and at peace. It was not as if I was the opposite of any of those qualities, it is just that now I am so much better than before. I began to really embrace change and love growth and life!

My health has also improved dramatically. I am sleeping well. Every day has become an adventure, filled with new learning. I harness the potential I felt was just below the surface of something that only now I can call "hardwired negative or self-limiting thought patterns and mindsets."

I have left those thoughts behind, in the past, over and done with.

A new me! A magnificent, empowering feeling!

A HOT TIP

Memorize The 12 Keys!

First, the act of memorization will force you to think about them more deeply and see the truth that lies within them.

Second, unless you can put your hand up and say, "My thought patterns, feelings, and emotions are perfect," you will find great benefit in reflecting on and using The 12 Keys as you go about your business and personal life. Use them to accelerate your growth and transformation. Become the businessperson you really want to be.

I have been working with them for years now, and I still find benefit and ever deeper layers of understanding and wisdom within them.

The more you work with them, the more they will work for you.

In Chapters 2–4 of *Get Wired for Success*, we will often return to The 12 Keys, as they are an integral and vital part of the journey ahead!

Now, start to use The 12 Keys in your life with Step 3 in online course!

Learning about The 12 Keys may have given you a new way of thinking, a greater understanding of your emotions, and a new way to see your world. Now visit the online course and complete the Step 3 worksheet. Record and consolidate your new ideas, and then turn thoughts into action!

Importantly, as you make your way through this book, you can revisit this online worksheet frequently to add new insights along the way.

KEEP IT SIMPLE

- The brain cell networks controlling your life are constantly changing.
- Neuroplastic change and the practice of "wiring your brain" are scientifically proven processes.
- Controlling this change to your own advantage is a skill called *self-directed neuroplasticity*.

The 12 Keys can change the way you think, feel, and see your world for the better, forever:

1. Hardwired brain cell networks are difficult to undo, but negative thought processes can be undone, and positive thought processes can be hardwired.

2. Searching for answers to negative thought patterns can reinforce them. Learn from the past, but let the past "be." Focus on positive outcomes.
3. New brain cell networks are easy to create!
4. Both the strength and number of connections between brain cells of a network increase with repetition, irrespective of whether the network is a positive or negative one. Practice works! Train and practice for success.
5. Change must be desired. The greater the desire, the greater the possibility of wiring or rewiring.
6. Remembering goals, strategies, and models guides positive wiring.
7. The greater the reward (or loss or failure), the bigger the neuroplastic change.
8. Mental rehearsal, meditation, and mindfulness can wire your brain in the same way as real-life experiences. This is the secret source of success.
9. Be aware: It is as easy to cause negative wiring in your brain as it is to create positive changes. Success requires discipline and persistence.
10. With every positive neuroplastic change, the brain takes the opportunity to weaken connections in negative brain cell networks.
11. Positive or negative emotions powerfully and instantly increase the number, strength, and durability of connections in a new or existing brain cell network.
12. You can wire your brain—and that of others—to feel positive emotions such as joy, celebration, gratitude, a sense of belonging, and happiness by creating new brain cell networks that release a cocktail of feel-good chemicals. Use the first 11 Keys to do this!

- Neuroplasticity will affect your success in business and life (positively or negatively), whether you like it or not.
- Knowledge of "*how to wire positively*" creates essential self-belief: "I can do it!"
- Knowledge of neuroplastic change *and the application of its principles to intentionally wire your brain* are vital if you want to create the business life of your dreams and live a life you *love.*

TRANSFORM YOUR COMMUNITY

I believe that, in the coming years, neuroscience and positive psychology will transform business management, leadership, and entrepreneurism. Excluding market and economic forces, and events such as disasters or pandemics that are beyond our control, how our business owners and leaders *think* will be the difference between success, mediocrity, and failure.

Do you think that creating an inspiring, vibrant, exciting, funny, compassionate, engaging, intensely interesting, progressive, and highly profitable business would lead to five-star service for your clients? I can tell you it does. Would that business make a difference to how your clients feel—cared for, welcomed, highly valued, or validated? Would they feel that here was a business exemplary in its behavior, that here was an example of how to *be* in this complex, fast-paced, and sometimes superficial world? The answer to each of these questions is a resounding YES!

Do you think it makes a difference to the people who work in a business—your team—and how they feel when they arrive at work? Would it make a difference to the level of their enjoyment while at work and to the standards by which they hold themselves true? Would it make a difference to their feelings of self-worth, how they feel when they go home at night, or even when their day is done? Again, the answer is YES!

Would that business's influence seep out into the broader community—that here, in this town, is a business that *cares*? Imagine if every business in your country was like the one I have just described above. What would it mean for your country, for its people, and for the world?

Would it make a difference to you? Knowing *how* to create such a business made an amazing difference to me!

THIS IS ABOUT YOU

At the start of Chapter 1, I asked you: Will you achieve your hopes and dreams? Do you have the personal power to make your dreams come true? Can you create a business life of purpose, passion, and success? Can you create a life of wealth, with the freedom to do what you want to do, and achieve real happiness, too?

Can you make an amazing difference to other people's lives? Think of the effect that would have.

Imagine living up to your amazing potential, unleashed into an extraordinary future. What a thought!

Other people have created such a life by *the way they think and feel and by the way they see their world*, and I too have learned how to think, feel, and see that way. Knowledge of the "*how*" in regard to how to change the way you think and feel and the way you see your world is a critical first step in creating an amazing business life, as well as an extraordinary personal life.

This relatively new and little-known science of neuroplasticity gives you the "*how*."

As a scientist, because the "how" is anchored in science, the result for me was a sudden and stunning self-belief: "I can do it! I can wire my brain for success!"

Now you can do it, too! Your brain is plastic! Direct your own neuroplastic change!

Once you know *how*, choices multiply, and the paths to follow become more expansive. The possibilities become endless, the potential unlimited, and the future exciting, almost breathlessly so. The creation of value is effortless. Wealth and success are abundant.[18]

This is your life! We have never had more opportunity than we have today, and the opportunities will be truly extraordinary and there for the taking in the coming years. You will change, and so will the world around you, rapidly so.

The "how" is up to you!

NEXT:

Now we better understand how you can effectively and positively turbocharge your thought patterns and emotions.

In *Chapter 2—Dream Big! Then Turn Your Vision into Results*, we will create *models* for both your business and personal lives to drive effective neuroplastic change. We will look at how to *direct* changes in your brain cell networks to your massive advantage and create the results you desire.

18 I will expand on these themes in Chapters 2–4.

CHAPTER 2:

Dream Big! Then Turn Your Vision into Results

So you can turn your dreams into reality!

"Whatever the mind of man can conceive and believe, it can achieve."

—NAPOLEON HILL, American author of personal success literature.

"Begin with the end in mind."

—STEVEN COVEY, American author, educator, and businessman.

IN CHAPTER 2

- How to create a crystal-clear vision of the future for your business—and your life.
- Seven rules for real results: Transform your dreams into reality!
- Four magnifying tools: Add speed and power to your vision.

THE BIG PICTURE

In the *Introduction*, I wrote that some business owners and entrepreneurs have created a life of wealth, freedom, and happiness by the way they think and feel and how they see their world. I learned to think, feel, and see the same way.

Chapter 1—How to Rewire Your Brain for Success gave you the science and knowledge about how you can rewire your brain to think, feel, and see that way, too. The "how" gives you the essential self-belief: "I can do it!"

Now in *Chapter 2—Dream Big! Then Turn Your Vision into Results*, we will look at how to give your thought processes and emotions specific and powerful direction by design so that you can see your world in the same way, too—your world in the future, even though you live in your current reality.

So, if Chapter 1 was about the "how," Chapter 2 is about the "where." It is about imagining your ideal version of your life. It is about designing a magnificent and wonderful future; it is the stuff that dreams are made of—and it is exciting!

Your vision should answer some of life's big questions: What do you want to achieve in your life and why? What feelings will you have when you have achieved your life's purpose? How will you know you have reached your goals? What do you want to be remembered for and by whom? Is there an end point to your vision, or do you want to keep on contributing until you can no longer do so? What is your life for?

WHAT IS A PERSONAL VISION?

Countless new-age authors, self-help gurus, and motivational speakers have encouraged us, inspired us, coached and coaxed us, and moved and motivated us to imagine and create a magnificent personal picture or "vision" of our own future. How right they are! Developing both a personal and a business vision is a critical step in creating your dream business and an amazing personal life.

A vision is an action plan for the future. It is an ideal image of how your world will be—how things will feel, look, sound, and even taste and smell—for you! Visions of the future are an essential part of our life. We could not do without them.

For example, a builder would not start to build a house without an architect's vision of its design. A pilot would not fly an airplane without a clear picture of the destination. A surgeon would not operate on a broken leg without a clear vision of the series of steps required to perform the operation, including the outcome.

In these three examples, we see the difference between a dream and a vision. However, the problem with dreams is this: They stay just wishful thinking, for example, "I wish I could make lots of money," "I wish my life were better," or "I wish I were fitter!" In comparison, a vision ends in results—the finished house, the plane landing at its destination, or the injured leg repaired.

When we are dealing with something as concrete as building a house, getting from A to B on a plane, or operating on a broken leg, the results are easy to see. Imagine an architect dreaming of a house, a pilot dreaming of their destination, or a surgeon dreaming of a repaired leg. How far would they get?

However, when we are creating our own personal vision, it can be difficult to create an ideal image. We live in our current reality, reached by virtue our genetics and our past. We cannot change either our genetics or our past.

We only live in our reality, so how can we escape it?

We escape our reality by using our amazing powers of creativity and imagination to see, hear, and feel things in our "mind's eye" before they occur. The vision you are about to create will be a picture of a "new you" in a personal and business world, with things the way YOU want them to be.

Why does this process work? It is reasonable for you to have doubts. A vision may sound too distant, unreal, unattainable, or even like feel-good nonsense! By creating a vision, you are creating a complex network of brain cell connections that define a model in your brain to guide your future actions—no less real than the models within the brain of the pilot, the architect, or the surgeon that guide their actions.

It is the Sixth Key to wiring your brain for success in action: Models guide positive wiring and rewiring.

The actions you take defined by this model subsequently create results—the results you want!

The difference in this case is that your model guides your actions toward your ideal future, but it is still a complex brain cell network, just like the architect's, the pilot's, or the surgeon's. If you further develop your vision with the other 11 Keys, your brain will be hardwired with an image of your future!

However, before you create your vision, there is a "but." As I suggested, dreams can stay as just wishful thinking. One guru made a fortune peddling the idea that if we imagine a dream and fiercely believe it will happen, then, hey presto, reality will manifest itself out of thin air.

Nothing could be further from the truth. Dreams without action remain just dreams.

We need *discipline* to convert our dreams into a vision and then into the action that delivers real results. Like any other endeavor in society—driving a car, playing a sport, or paying bills—we need rules to define the discipline required for the task ahead.

To that end, I use Seven Rules for Real Results. They turn my dreams into a vision and then into the amazing business and personal results I achieve. Now, you can do it, too.

THE FIRST SIX RULES FOR RESULTS

The First Rule: Dream big! There are no limits!

Wow! This is at once extraordinary and liberating. If other people can create amazing lives and you know the "how," then you can, too!

This is how you leave behind your self-imposed limitations, mindsets, inhibitions, anxieties, and even your fears. This is how you leave behind other people's opinions and the "scripts" of what to say and what to think, given to you by your parents, teachers, friends, past employers, and those who would like to see you fail. Leave behind their expectations. Leave behind society's norms. Leave normal behind!

This is how you put the past with all its negatives behind you.

This is how you take the past with all its positives with you.

This is how you realize your amazing potential, even though you may not recognize just how great you are now. The fact that you are reading these words now, driven by *something*, tells you there is more in your life. The quivering emotion you feel when you imagine an amazing future is *something*. It is your untapped potential, lying just below the surface of all those dull and stifling negatives, waiting for you in your subconscious, waiting for you to wake up.

Wake up! The past is over.

Think of the possibilities! Dare to be great; dare to be extraordinary! Set a magnificent example for your family, business, friends, and community.

This is how you reach for the sky!

The Second Rule: You must believe you can do it

Your vision must be totally believable to you (even if only to you alone), and you must totally believe in it. We all know that if we stand on a ledge, jump, and flap our arms, we cannot fly. No amount of vision or belief will achieve unassisted flight. So it is with your vision.

In the *Introduction*, I wrote that this process is like building a skyscraper. You need exceptionally strong foundations if you are going to reach for the sky. *Self-belief is a foundation stone for your path to the sky.*

Creating an amazing business and personal life is all about seeking information, acquiring and applying knowledge, and gaining wisdom. You do not need good looks; you do not need rich parents; and you do not need a university education. Neither is it about who you are, your past, or your gender or age. Rather, it has everything to do with *what you seek, what you know, what you do, and when, where, how, and why you do it.* You must believe this from deep inside and with all your heart.

So, the fact that you must believe in your vision does not set limitations on it. Quite the reverse: Believe that your potential is unlimited!

The Third Rule: You must create both a business and a personal vision

These two parts of your life are tightly linked: Your business life is the vehicle for your personal journey. In the same way a car is simply a vehicle for getting from A to B, your business life gets you from where you are now to a desirable point in the future. It produces a financial result that affects your personal life—for better or worse. It takes up time and energy that also affect your personal life, and its success (or mediocrity or failure) will consequently influence your happiness, self-esteem, and confidence—in turn affecting your personal life.

That said, your personal life also influences your business life, including your *performance* within your business, therefore contributing to its success or failure.

You cannot separate one from the other. Instead, create an amazing version for both. Each can and should augment the other in wonderful synergy.

The Fourth Rule: You must take specific actions to achieve your goals

Start to create your dream business life *now*—even though some of the steps required are not obvious to you.

Steve Jobs, the cofounder of Apple, made a speech about predicting the future to the commencement class of 2005 at Stanford University:

> "You can't connect the dots looking forward; you can only connect them looking backward. So, you have to trust that the dots will somehow connect in your future. You have to trust in something—your gut, destiny, life, karma, whatever. This approach has never let me down, and it has made all the difference in my life."[19]

Conversely, we all know people who have great dreams and plenty of ideas but never act on them, so their ideas stay just that: dreams. Author James Thurber's fictional character Walter Mitty is the most famous example of living a vivid fantasy life but being an ineffectual dreamer.

I found that once I fully understood the creative process, taking the required specific actions to create a dream business and an amazing life was simple, easy, and intensely rewarding. It is not hard work at all; in fact, it is a joy.

This book and the companion online course are carefully designed to turn your work into successful action so that you are free to create *your* path to the business life of your dreams, living a life you love.

The Fifth Rule: Your life partner and your business partner(s) must be involved in vision planning

---

19 Steve Jobs, "Steve Jobs' 2005 Stanford Commencement Address," YouTube video, 15:04, March 7, 2008, https://youtube/UF8uR6Z6KLc.

Ideally, your life partner and business partners (if you have them) should create their own visions, and they should be completely aligned with yours. The less the alignment, the more problems in the future—your problems!

Imagine if you dreamed of creating wealth to experience incredible holidays in the most romantic cities of the world, and your life partner wanted to create wealth to help support an orphanage in a Third-World country, but neither of you wanted to support one another's dreams. Your results would be so much less due to the conflict of interest and negative tension it would create. Of course, you could do both and be richer for the experiences—the combination could be an amazing adventure.

Two minds are better than one—if they are heading in the same direction! Alignment is so important.

The Sixth Rule: You must use your new power only for good

This process—together with some of the other techniques in this book—can give you amazing, new personal power. Promise to use it only for good. Promise!

Now, create a crystal-clear vision of your dream future

Find a place that is peaceful and quiet, where you can be alone with your thoughts. A darkened room with a comfortable chair is fine, or you may prefer a garden or a beach for solitude.

Then, close your eyes and imagine this: I have a time machine, and I invite you to take a trip in it. In your mind's eye, using the power of your imagination, climb inside my machine, start it up, and set your destination to the future when you have created the *business* life of your dreams.

Next, take a deep breath, relax, and feel at peace. Set your mind free. Open yourself to the possibilities of your amazing potential and take that jump in time!

Project into the future with your imagination, unrestrained, excited, and bold. See yourself being the person required to make your dreams come true. What type of person would you have to be? What character traits would you exhibit as a businessperson, a manager, a leader, a business owner, or an entrepreneur?

How would you look? How would you speak? How would you feel when you achieve your goals?

Think deeply about the answers to these questions. Take your time... Develop a crystal-clear image of yourself in your mind's eye, living the business life of your dreams.

Describe that image of yourself in words. Your description creates a **vision statement of your business *life***[20]—the person you will *be* in order to build the business life of your dreams.

Start to create the business life of your dreams by taking Step 4 in the online course now!

Next, design a crystal-clear vision for your personal life.

20 This statement is for your private use. It is not to be confused with a **vision statement of your *business*.** This is developed together with your team for your investors and clients. It tells all stakeholders where your business is heading in the future and is a public statement for all to see.

Now, set the destination of my time machine to a point in time when you have achieved all your personal dreams and goals residing deep within you and are living a life you love. Take that jump in time!

What do you want to achieve? What do you want to own? Where do you want to go? What feelings do you want to feel? What relationships would you like? What are your deepest desires? Most importantly, who do you want to be?

Go for it! Really unleash your extraordinary potential and imagination and go for it! Remember, there are no limits, but it must be believable to you. See it all in your mind's eye.

Now, develop a crystal-clear image of yourself loving the life you live. Again, describe the image in words. Your description creates a **vision statement of your personal life**, describing the ideal person you will *be*, living a life you love.

Start to create a life you *love* by taking Step 5 in the online course now.

The two statements are your vision of an amazing future—your dream business life and YOU loving the life you live!

Steps 1 and 2 in the online course describe your Point A—where you are now—and my time machine transports you to your Point B—where you really want to *be*. The two vision statements describe and detail your Point B.

You may experience a range of emotions you have never felt before. This can be extremely liberating, uplifting, and inspiring. Of course, we are just using the time machine as a tool to free you from your past, to unleash your imagination, and to create a future you.

Armed with the knowledge of how to wire and rewire your brain, your vision creates both a model for future action and vital, positive self-belief. This is a critical step on the path to creating a life of wealth, freedom, and happiness.

If you design your vision correctly, it will be a series of many wonderful, gigantic, exciting, colorful, multisensory, and complex images, each defined by a network of strong and enduring brain cell connections within your mind. This will function as a detailed instruction manual to confidently align and guide your future actions.

Make the vision statement of your business life for yourself, but also make it for your investors, team, clients, and suppliers.

Make the vision statement of your personal life for yourself, and for your family, friends, and community—even make it for humanity!

Make it BIG! Make it magnificent.

HOW YOUR VISION DETERMINES YOUR FUTURE

With the power of your imagination and creativity, you are literally designing your future. This important section integrates your vision with both the science of neuroplasticity and how a successful vision creates your reality and results. Let me explain by way of example.

When I took over sole ownership of my business, I had a definite business vision in mind. When I look back, I am surprised at just how limited it was, but I had no knowledge of many of the powerful concepts within this book or how to use them.

I had witnessed firsthand the terribly negative effects of a dysfunctional business partnership. Interpersonal conflict and lack of aligned visions were the causes. Imagine watching business partners shouting at one another within earshot of the team. Imagine observing clients being treated like a nuisance. Imagine clients being shouted *at*.

I felt continually on edge, anxious, and nervous. When would the next horrendous drama unfold? When next would my heart race? When next would my stomach churn? What a way to do business!

If one word could have described my dream for my future business, it would have been the exact opposite of conflict: harmony—between myself and the team members and between the team and our clients. If I took on a prospective business partner, then a harmonious partnership would be foremost in my mind.

Peace, calmness, and simplicity are other words that come to mind. If you had asked me at the time how I would have liked things to feel, I would have replied: "As peaceful as the inside of a Buddhist temple."

This is how my vision unfolded and how it relates back to the science of neuroplasticity:

- At some point, one of my team members asked me if I wanted attractive candles burning in the office, as they would create a warm, tranquil, and relaxing atmosphere—an idea consistent with my vision. "That is a good idea! Yes, please," I replied.
- On another occasion, a salesman asked me if I would like beautiful, artificial flowers in the reception, changed regularly and color-matched with my business. My answer was automatic. I did not consider the cost, even though it was minor—another idea consistent with my vision.
- Our business logos were initially designed in an Oriental style with a calligraphy pen. They are simple, elegant, and professional. This time, *design* is consistent with my vision. Now, the logos are displayed on the massive smoked glass, floor-to-ceiling windows at the entrance to my business.

- I used the words "harmony" and "empathy" in selecting and training team members. These key words are used in job adverts, interviews, and performance reviews and are two of our six business values—prominently on display in team areas. These were *language and business standards* consistent with my vision.

All these decisions were made to evoke harmony, peace, and simplicity.[21] The Buddhist temple was a powerful visual and an emotional metaphor for what I wanted.

Years later, I walked into my reception area. The candles still burned invitingly; the flowers looked stunning; the logos were amazing. It was uncluttered and clean, suggesting thoughtful organization. But here is the thing: Two team members were engaged in a warm and friendly discussion, another was comforting a worried client, and a fourth was on the phone, welcoming a new client. Harmony and empathy! I felt relaxed, confident, and happy—what a contrast to those early days!

My dream was unfolding before my eyes in real time. I took a deep breath of quiet satisfaction. Inside, my heart was bursting with pride.

They say that pride is one of the seven deadly sins. Some types of pride may be so, but not pride in accomplishment.[22] My pride in such a wonderful transformation—in both my business and myself—was a magnificent feeling.

---

21 You will find that concepts such as harmony, peace, and simplicity are recurring themes throughout this book. Why are they so important? Through them, you find clarity, and clarity enables concentration and discernment. These are key skills required to build your ideal business life. Initially, I had no idea how to achieve this vision of harmony, peace, and simplicity, but as Steve Jobs wisely said, the future is often hidden from us.

22 We will look at the key role of accomplishment in generating positive thoughts and emotions in the *Conclusion*.

My early and limited dream for my business was now reality. Strong brain cell networks associated with a desire for concepts such as harmony governed my subsequent decision-making. Some decisions (like the candles and flowers) were made at the time without thinking about the end effect. They just seemed like no-brainers. Other decisions were seriously and carefully thought out to bring about the same result, such as the logos and the words used to select and manage my team members.

This is the sequence: Dream big, create a crystal-clear vision, hardwire it using The 12 Keys, take actions perfectly aligned to your vision, and then get the results YOU want! Now, decisions regarding the presentation of my reception area and offices are easy. I am confident in what I am trying to achieve. My vision is always foremost in my mind.

Imagine taking my early limited vision of a business culture with harmony and adding more expansive elements in other areas such as strategy, finance, marketing, customer service, and so on. Clear vision in each of these areas would create decisive and intelligent actions, in turn creating the desired results—or at least close to them.

Note that both business and personal visions can be as much about emotions, feelings, sounds, and even smells and tastes as they can about the visual elements of our life. Think about the importance of the sense of smell to a business such as a medical or dental clinic or the sense of taste to a restaurant or café. You should consider all the senses in your vision of your business.

So, although the past (and the lessons we learn from it) is important, sometimes crucially so, your vision is about saying goodbye to the past. It is about designing your future and not being controlled by your past or by external factors that tend to "bounce you around." Undirected, reactive decision-making leads to chaos. What we want is discernment and organization, not chaos.

Creating your vision is about using your imagination to create a "new, highly focused, highly directed version of yourself."

You are different to me. As you read these words, I cannot hear your story and how you got to this point in your life, but I can ask you, "What do you have to change and how do you develop to bring your vision to reality, to be the manager, leader, business owner, or entrepreneur creating your dream business life?"

If you reply, "There is nothing to change. I am completely and totally satisfied with my business and personal life. I am living the dream," then congratulations! That is wonderful!

For those of you who dream of untapped business and personal potential, or who have self-imposed barriers preventing you from reaching your full potential, then let your imagination run free. Imagine yourself living your dream business life. What would it look like? What sounds and voices would you hear? How would it feel? Imagine living a life you love!

Take your time. This is your future!

If you are successful in creating the business life of your dreams and living a life you love, you will have changed into a "new you." Neuroplasticity will guarantee that.

The Seventh Rule for Real Results: Be as if your business and personal visions are your reality NOW

My seventh rule is the one guaranteed, must-do strategy to go from dreams to vision to results.

Once I had my knowledge of neuroplasticity and a crystal-clear vision for both my business and personal life, I had the solid self-belief of knowing the "how" and my destination (the "where"). I asked myself a simple question: When do I start to change? I immediately realized that the obvious answer was NOW.

Mediocrity, with anxiety as a constant companion, was no longer an option. I can remember that moment clearly—feeling a shiver of fear and a quiver of excitement. Why fear? From my low point of poor self-esteem, anxiety, insomnia, and being burned out, I realized what a massive change I would have to make to my thinking and emotions.

Could I make such a huge transformation? I did not have all the information I have at my fingertips now. I was still living in ignorance.

It was scary. What if I did not achieve my vision? I knew I would die a frustrated, old man, and I did not want that!

I wanted to live my life to the fullest. I wanted to experience the potential that was always there, just below the surface. I wanted to be free of limitations, doubt, anxiety, the past, other people, and to give to others as I wanted, on my own terms, in my time, independent and interdependent, to be amazing and inspiring.

I wanted total freedom to be *me*! Yes, it was scary.

Why excitement? My vision for my business and personal lives lit a fire within me that still burns fiercely today.

Why is the Seventh Rule for Real Results so important? Consistently positive neuroplastic change—wiring your brain for success—can only take place in the present.

If you wait for it to take place, wish for it to take place, think you deserve it to take place, or think that it will just fall out of the sky into your brain at some future date, it will not happen. It will not happen until you start to be as if the future were here now—not tomorrow, not next year, or in five years. NOW!

Why now? Your negative thought processes and emotions are hardwired. The only way to weaken them is to let them *be*, to create new, positive thoughts and emotions, and to strengthen existing positive ones. *That takes time—sometimes a lot of time!*

Why now? If you do not start to change now, you will continue how you are now—with a life as undirected and as reactive as it is—and only you know what your life is really like. Chaos can reign! I know—I have been there.

Why now? If you do not change now, you will not be able to lead the other people in your life properly and appropriately to help you achieve your vision. So, by definition, you will not achieve your vision. YOU come first—not the other way around. People will not change serendipitously just to suit you. I tried that way for eight long years. It was a heart-aching disaster.

Why now? It is *so* exciting to create the business life of your dreams and live a life you love. It easily beats mediocrity or failure!

Why now? How much time do you need? How much time have you got? I do not know the answer to those two important questions, and neither do you.

Why does it *seem* so hard?

Our weaknesses, negative or self-limiting mindsets, thought patterns, beliefs, behaviors, attitudes, anxieties, fears, and (even for some) obsessions or addictions are all hardwired—so much so that we cannot even recognize them for what they are.

They are an integral part of us, of who we are now—the product of our genetics and our environment, everything that has happened to us in our day-to-day lives until NOW. They are both conscious and subconscious, influencing and controlling, prodding, poking, and seducing us to take the actions we take now to get the results we get now.

Remember this: Other people have created a life of wealth, with the freedom to do what they want to do, and with real and enduring happiness, too. They have built a business life of purpose, passion, and amazing success *by the way they think, the way they feel, and the way they see their world—by the way they are wired.*

So, if you want to feel on fire with motivation and purpose in your life, start having a *compelling purpose* NOW, even if you are totally burned out and want to give it all away. Developing and acting upon compelling, challenging, and intensely rewarding purpose brings motivation and enthusiasm to your life.

Why does having a compelling purpose lead to success? It creates work that is absorbing, interesting, and challenging. Work starts to *flow*, and so much more gets done. Other people start to follow you. Purpose builds teamwork and cohesion, a positive culture is created, and the fun starts. Doing rewarding work is a reward in itself.

By developing the beliefs, attitudes, and behaviors of people who live with compelling purpose, you will take the actions that create a very purposeful life. If you start having compelling purpose, you will start getting the results of people who have compelling purpose in their life right now.

Slowly, the brain cell networks controlling feelings of purpose will wire and rewire themselves to generate feelings and thoughts of purpose (we will look at developing a compelling purpose in *Chapter 3—On Fire with Purpose, Power, and Passion*).

If you want *red-hot passion* to be a core part of your life, you must start being passionate NOW. Start having the attitudes and beliefs of a passionate person. Have the behaviors and take the actions of a passionate person, even if your current life is dull and boring and you are screaming to escape it.

Why is red-hot passion in your business life so effective? You cannot create a dream business life and an amazing personal life by yourself. Passion is essential to lead other people—primarily, your investors, team, clients, and suppliers—in the creation of your vision.

By developing the beliefs, attitudes, and behavior of passionate people, you will take the actions that make passionate people deeply passionate. If you start being passionate, you will start doing things that give you that great feeling. You will follow your heart. You will begin to *feel* with more emotion, excitement, and enthusiasm. *You will start to get the results of passionate people.*

As with feelings of purpose, brain cell networks controlling feelings of *passion* will wire and rewire themselves to generate feelings and thoughts of passion. We will look at ways to develop red-hot passion in both Chapters 3 and 4.

If you want to be *amazingly successful* in life, you have to start having the attitude of someone who is successful; you have to have the beliefs of a successful person and the behaviors of a successful person right NOW, even if your self-esteem is at rock bottom and you think you are a failure. Even if you have made massive mistakes, carry yourself as if you were successful beyond your wildest dreams.

Why? Other people will pick up on it. Eventually, they will start to follow you, and suddenly there is a heady atmosphere of success. Things will start to click. There is a buzz, and it is intoxicating.

By developing the beliefs, attitudes, and behavior of successful people, you will take their actions. If you start doing the things that make people extraordinarily successful, then you will start to get the amazing results they get right now. Again, brain cell networks controlling feelings of *success* will wire and rewire themselves to generate feelings and thoughts of success.

If you want *freedom* in your life—to do what you want and not be trapped by your circumstances—you have to start having the attitude of someone who *is* free; you have to start developing the beliefs of a free person and the behaviors of a free person right NOW, even if you feel as if you are hopelessly trapped in a living nightmare. Act free in your world right NOW!

Why? You cannot own a dream business in which you display emotions consistent with the feelings of being trapped. Your negative emotions will be toxic in the workplace.

Conversely, positive emotions are synonymous with success. Walk into any business and check out the atmosphere. Is it negative or positive, or is it toxic or successful? You must continually demonstrate positive emotions in your business. Other people—primarily your team and clients—will pick up on your vibes and follow.

By developing the beliefs, attitudes, and behavior of someone who is free, you will take the actions that ultimately lead to your freedom. If you start doing the things that ultimately set you free, then you will start to get the freedom you so deeply desire—and, yes, brain cell networks controlling feelings of *freedom* will wire and rewire themselves to generate feelings and thoughts of freedom.

If you want to be *wealthy*, start having the beliefs and attitudes of a wealthy person. You must start behaving as if you were wealthy, even if you are broke or deep in debt. Why? Remember, you need a financially successful business life if it is to be the vehicle of an amazing personal life. Financially successful businesses can afford to employ high-quality people. They can invest in new equipment to keep up with their changing environments. They can market themselves appropriately, and so on. Success leads to even more success.

Does that mean you can spend money like a wealthy person now and start living the high life? No. Start investing and saving like a wealthy person, and only spend money on life's luxuries when you can afford to do so—*without jeopardizing your accumulating wealth.*

Brain cell networks controlling the beliefs, attitudes, and behaviors required for wealth will wire and rewire themselves to generate the beliefs, attitudes, and behavior of wealthy people. You will take the actions that make wealthy people very wealthy. If you start doing this, you will begin to get the results of wealthy people. It is what they do and refrain from doing that makes them wealthy. It is not their luck, genetics, or environment—it is what they do.

If you want *real, enduring, and deeply satisfying happiness* in your life, you must start having the attitude of someone who is genuinely happy. You must start developing the beliefs of a happy person, including their behaviors, right NOW. Even if your emotions are low and enduring and deeply satisfying happiness seems like a dream at the moment,[23] carry yourself as if you were the happiest person in the world right NOW!

Why? You cannot own a dream business in which you are consistently or even occasionally unhappy. Again, your negative emotions will be toxic in the workplace. To repeat, positive emotions are synonymous with success, and you must continually

---

23 If you are experiencing anxiety, depression, suicidal thoughts, obsessions, or addictions, seek the help of a qualified health professional now.

demonstrate positive emotions in your business, and other people will follow. If you want engaged and motivated team members and clients, set the example.

> ***Remember, the Twelfth Key states that you can rewire your brain to feel positive emotions such as joy, celebration, gratitude, a sense of belonging, and happiness by creating new brain cell networks that release a cocktail of feel-good chemicals.***

By developing the beliefs, attitudes, and behavior of happy people, you will take the actions that make happy people so incredibly happy! Happiness is the result of what you *do* and not a default state just because you are human.

If you start doing the things that give people genuine and enduring happiness, then you will start to experience the happiness they feel right now. Brain cell networks will release feel-good chemicals that start flooding your brain, stimulating pleasure centers, and generating increasingly better feelings. We will look at ways to feel deeper and longer-lasting feelings of happiness in *Chapter 4—How to Master Your Thoughts and Emotions.*

In summary, what thoughts and feelings are important to you? Wire your brain for __________! (Fill in this space with the outcomes you desire!) Start living the life you want and so richly deserve NOW!

Reflect for a moment… As you can imagine, trying to adopt all these personas at once does not *seem* easy, particularly as much of your current persona is hardwired. Aligning your attitudes, beliefs, behavior, and actions with your vision can *seem* overwhelming.

I can relate. When I was overwhelmed with my negative thoughts and emotions and limited by my ignorance, it was extremely hard. I did not have The 12 Keys and how they relate to vision, purpose, and passion. I did not know how to master my own thoughts and emotions. A lot of pieces from the jigsaw puzzle were missing.

Learn from your past, but let it be. Start with your current reality. *Change begins with a first step.*

Using The 12 Keys to wire your brain for success, start to be the person of your vision now. Your plastic brain will start to change, imperceptibly at first, but it will change.

Will you "get it" right immediately? Will you become that person of your future vision now? No. That is why both vision and practice are so important.

> ***The Fourth Key to wire your brain for success tells us that practice increases the number, strength, and durability of (in this case, positive) brain cell connections.***

Head in the direction of your vision. Practice is the process of reflecting on your progress, adjusting, and continuing your journey.

This is the expanded sequence: Dream big, create a crystal-clear vision, hardwire it using the 12 Keys, take actions perfectly aligned to your vision, and get results—then, reflect, adjust, and continue… make a little progress… then reflect, adjust, and continue again. Slowly, your plastic brain changes, brain cell by brain cell, connection by connection, one at a time, millions upon millions in a day, positive connections becoming stronger, until it starts getting easier and you start noticing the positive difference.

Now, I have a life of passion, purpose, and amazing business success. I have created the business life of my dreams, and I love the life I live. It has been created by the way I learned to think, feel, and see my world and by my beliefs, behaviors, attitudes, and actions. That is when the real fun starts!

Before we continue, there may be some part of you questioning whether this is selfish—designing a personal and business life just to suit me?

The answer is yes and no. You can have (and should and deserve to have) the best of both worlds—living a wonderful life *and* being of massive value to others. In fact, you will struggle to bring your dreams to reality unless you are of significant value to others. You will see this concept slowly unfold as you create the business life of your dreams and live a life you love.

This is first and foremost about YOU, and then through you and because of your actions, it is about the other people in your life and community. It is a win-win situation—synchronous, synergistic, integrated, holistic, and harmonious.

Because wiring my brain for success *seemed* so difficult, I developed four tools to make the process easier and quicker.

FOUR MAGNIFYING TOOLS TO ADD POWER AND SPEED TO YOUR VISION

The First Tool: Discover the power of clarity

World-famous motivational speaker Tony Robbins uses the phrase "clarity is power," and he is so right.

Imagine driving a car in heavy, early-morning fog. You drive slowly and very carefully. Your eyes are glued to the road just in front of the car, and you feel tense with concentration. Slowly, the fog begins to clear, and you drive faster. Suddenly, the fog disperses, and you burst into sunlight. The way forward is easy!

You accelerate, relax, and begin to enjoy the journey, taking in the scenery.

So it is with your journey to a wonderful and magnificent future. If your vision is foggy and lacking detail, your journey is more difficult, slower, and stressful. If your vision begins to clarify, your journey is easier, quicker, and more enjoyable. If your vision is crystal clear with incredible detail, your journey is the easiest, quickest, and most enjoyable. It is important to *enjoy the journey* as well as the destination!

Why does clarity give you power? Clarity adds detail. The more detail, the greater the number of steps in your personal and business vision statements. The "distance" to your vision remains the same, so *the more steps you create, the smaller the distance between each step*. As each step becomes smaller, it becomes easier to achieve.

You make progress more easily. Progress builds confidence, and confidence gives you power!

> ***Remember the Second Key. Positive brain cell networks become even stronger if we focus on positive outcomes.***

Each step toward your vision is a positive outcome. With each step achieved, the brain cell networks that enable your imagination and describe your vision become stronger and more enduring.

Many people fail to achieve their vision because the size of the task overwhelms them. They fail to start or become discouraged early in their journey, and their vision remains just a dream. By breaking your vision down into smaller steps, the work ahead becomes exciting and achievable.

ANOTHER HOT TIP

Your current situation in business and life is your Point A, and your dream destination is your Point B. You only need to identify the very first step on that journey from A to B. Focus on it! Take it, and once you have taken that first step, reassess, and then the next step will become clear.

You do not need all the steps between Point A and Point B. In fact, some of them will be hidden from you, as twists of fate will take you down paths that you cannot imagine now. The key is to take that first step!

Now is the *perfect* time to get clarity and power by completing Steps 6 and 7 in the online course.

The Second Tool: Make the journey to your vision really rewarding!

My seventh rule, being as if your business and personal visions were reality now, can be challenging and may take a long time to achieve. Now, we have just added more steps to your vision with added detail. It can begin to sound too difficult, so it is important that you reward yourself along the way.

We have all heard stories of people who have thrown themselves into their business at the cost of their personal lives or sacrificed business success for a certain lifestyle. You can get buried in your business life or in your personal life and never make real progress. I have experienced both scenarios.

The two lives should complement one another in perfect synergy. Your business life should be the vehicle for an amazing journey in your personal life and not a weight upon you and other important people in your life. Your personal life should be a celebration of what you have achieved in your business life.

Given the challenge, it is critical to have a genuinely great time on your journey and reward yourself regularly. It is not so much about the destination as it is about the journey. *Without the journey, there is no destination.*

If you want to be wealthy, how do you reward yourself and not *jeopardize* your financial position at the same time? Use your imagination—there are many rewards that cost almost nothing. Take time to relax and listen to your favorite music, watch movies, walk in your favorite outdoor areas, run, go to the gym or play sport, eat your favorite food slowly and really enjoy it, and, of course, spend time with family or friends. What do *you* really enjoy?

While these may be everyday events and may be taken for granted, how often do we take the time to really savor them… to refresh our spirits and nourish our souls? How often do we take time to reflect that these are real and genuine rewards of life?

Here is an important point: Every time you stop and reward yourself, think of the progress you have made on the way to your vision. Enjoy the feeling of achievement. One positive step, no matter how small, is one step closer to your vision.

As you begin to accumulate wealth, you can afford greater rewards for the positive progress you have made. Suddenly, you can afford an evening out, then a weekend away, and then a special holiday. Suddenly, you can afford material possessions if you want them, small at first but more expensive with time.

However, each reward must not jeopardize your increasingly strong financial position.

There is good reason for including rewards in this process.

Remember the Seventh Key: Rewards drive positive wiring and rewiring.

You will enhance positive wiring when you reward yourself in a timely and appropriate manner during your journey. Importantly, you will begin to consciously link the reward to the actual progress you have made.

Also remember the Twelfth Key: You can rewire your brain to feel positive emotions such as celebration by creating new brain cell networks that release a cocktail of feel-good chemicals.

In the case of reward, the feel-good chemical is called dopamine. Every time you set a new goal that is within your reach, dopamine is released, causing a feeling of pleasure, as you anticipate the reward. Another flood of dopamine is released in *celebration* when you make the goal and get the reward. More pleasure!

Then, you set the next step in your vision. *Each step in your vision is one step on your path to success.* Reward reinforces positive wiring!

Now complete Steps 8 and 9 in the online course. Make them fun. Enjoy!

The Third Tool: Add "wow" goals to boost your vision

Remember, the Fifth Key states that change must be desired. The greater the desire, the greater the possibility of rewiring.

When you are fired up for change, when your brain is desperately desiring an amazing amount of pleasure, and when motivation and engagement are high, the brain releases chemicals that accelerate the process of positive rewiring. Change is easy.

So, let us hijack our brain's amazing ability for positive wiring to our own advantage: Add some enormous, totally awesome, and incredibly rewarding goals to your vision and business statements—goals so good that when you achieve them, you will be left in stunned disbelief saying, "Wow!"

While vision statements are how things will look, sound, and be at some point in the future, a goal is a specific and concrete achievement. *Wow* goals are your most magnificent rewards—ones you enjoy when you reach major points on your journey. You deserve your *wow* goals because of all the good work you have done. Once you "kick a *wow* goal," you can plan for the next one—or more than one!

My own personal *wow* goals (and I have several) are held deep within me. I've "pulled them out" and imagined *having reached them* when the going has gotten really tough, when my life has lacked purpose or passion, when success has eluded me, when I have felt totally trapped by my circumstances, when wealth has seemed a distant dream, or when happiness has been a fleeting emotion in a dull and gray world.

Wow goals motivate you to take the next step forward, a quick return to positive wiring, positive thought processes and emotions, positive actions, and positive results. *Wow* goals also have the benefit of stretching and expanding your vision. If a *wow* goal is what you really want, deep down inside, then it can turn your vision from good to totally awesome.

Remember, a *wow* goal is for YOU and only you. It should give you immense pleasure, but it may also bring pleasure to the other important people in your life. A *wow* goal can be inclusive of other people, but you are the most important recipient.

Now, take Step 10 in the online course (my favorite Step!)
It is important to revisit this step as you make your way through the rest of the course.

The Fourth Tool: Make your vision of an ideal future endure

Your business and personal visions are "big-picture" statements—holistic images of how you would like your life to be. They are difficult to carry around with you daily as life often demands your full attention. Reading them on a regular basis provides clear direction for your behavior and actions.

Remember the Fourth Key: The number, strength, and durability of connections in a brain cell network—in this case, that of your vision—increases with repetition.

Revisiting and revising your vision etches and embeds your wonderful future deeper into your mind, almost as if it were reality, making it easier for you to make consistent decisions in the direction of your vision—just as I did in the design of the reception area in my business and in thousands of other decisions in my business life. It also allows you to adjust your vision as time moves on and as your environment changes. These are not static documents. Edit them regularly!

Further clarity can also be added as the future unfolds. Extra clarity boosts power!

Initially, I revisited my vision statements daily as I continually improved them, and then once a week. Now, I fine-tune them once a month. It is still important to do as it is easy to stray from the direction of your vision and lapse back into undirected behavior.

I personally found that although my vision statements became bigger, bolder, and clearer, little of the original intent changed. The first version provided a great framework for increasingly better versions. Importantly, when you consistently make decisions aligned with your vision, you start living your dream!

Make sure your vision endures with Step 11 in the online course!

Revisit this step as you make your way through the course.

Once you have hardwired your crystal-clear, multisensory, three-dimensional, and *exciting* vision of your ideal future, you will have a route map for the journey ahead. It will be stored as a complex and enduring network of brain cells in your mind, instantly ready for recall, waiting to guide your actions. Congratulations!

KEEP IT SIMPLE

- Business and personal visions are plans—ideal images of how things will feel, look, sound, smell, taste, and be in the future.
- Your business and personal vision statements are *models* that will drive your future beliefs, behaviors, attitudes, and actions—*they are the Sixth Key to rewiring your brain in action.*
- Use my "time machine" to create a powerful vision of the business life of your dreams.
- Use the time machine again to create a powerful personal vision of *you* as you really want to *be*—free from the influence of the negatives of your past, living out your amazing potential, living a life you *love*.
- Write down your visions on a computer, laptop, or tablet—this increases the likelihood they will become reality.
- The two vision statements create positive self-belief.
- Share your business and personal vision statements with your business and life partners. Ideally, align your vision statements with theirs. They are statements of how you desire your world to be. They should guide and govern your beliefs, attitudes, behaviors, and actions.
- Add clarity to your vision statements by adding greater detail. Clarity gives you confidence and power.
- The most important step in your journey is the first one. Others will be revealed in time. Make a start!
- With progress, reward yourself to reinforce positive wiring, and get a bonus boost of the feel-good chemical dopamine!
- Meaningful rewards can cost nothing.
- Add *wow* goals to your vision to create strong brain cell networks of desire.

- *Use the Fourth Key to wire your brain for success: Practice makes your positive brain cell connections more numerous, stronger, and more enduring!* Revisit and edit your vision statements on a regular basis to embed your vision for enduring success.

TRANSFORM YOUR COMMUNITY

Once you know the businessperson you want to be, your confidence will rise. Your influence will expand, and you will begin to glimpse your greatness!

Above all, our businesses need businesspeople with a clear and inclusive vision for their investors, teams, suppliers, and clients. You can lift your team up from the mundane reality of day-to-day work. You can inspire motivation and cohesion, uniting your team in common goals of real purpose. You can tell clients not only where your business is heading but also what motivates it daily as it strives to reach a lofty destination.

You will hear comments like: “It is a great place to work” and “a great place to do business.” Compliments will become commonplace.

Your influence as a businessperson can also filter into the community as an example of progression, dynamic and positive change, stability, and success. Imagine a community in which every businessperson has a clear picture of his or her vital role in helping their business achieve extraordinary goals. That is the challenge we have ahead of ourselves, one which demands our imagination, concentration, discernment, and commitment. It demands your vision!

IT IS ABOUT YOU

Yes, this really is about you. Only you can tap into your deepest feelings and desires to create a magnificent future. If you really want a life of wealth, freedom, and happiness, the opportunity is before you. Use my journey as an example. This is YOUR journey, and only you can take it.

It comes down to what you chose to do every moment for the rest of your life. You can wallow in failure or stumble through life in mediocrity—"bounced around" by circumstances beyond your control—or choose a magnificent future.

Are there any guarantees that you will realize your personal vision? No. External factors that affect our businesses are beyond our control, as are some events in our personal lives. We cannot predict the future. Indeed, we seem to be living in increasingly unpredictable and uncertain times. The pace of change is accelerating. We must learn to live with change, ambiguity, and uncertainty.

In addition, as we move through life, our vision can change, too.

The good news is that by visualizing your ideal life now and progressing toward it, your vision will expand, more opportunities will reveal themselves, and more rewards will follow—many of them non-financial. Achievements built on big visions, positive emotions, and great relationships will reward you in ways you cannot imagine. It is all about the journey.

What I can say is this: Tip the odds massively in your favor. Adopt the beliefs, behaviors, attitudes, and actions of people who have a compelling purpose, red-hot passion, and amazing business success. Think the way they think; feel the emotions they feel; see your world the way they see theirs. Create *your version* of wealth, freedom and, yes, real happiness.

The 12 Keys to Wire Your Brain for Success—based on science—give you the "how" and, in turn, the essential self-belief to say, "I can do it!" Your crystal-clear vision gives you your destination. Take the first step toward it!

Enjoy the wonderful journey on your way to a magnificent future.

It starts now!

NEXT:

Once you have created your detailed personal and business vision statements and a list of rewards and *wow* goals, you now know where you want to go. In my case, with my limited business vision, the path ahead was foggy, but the fog soon began to clear—I knew where I wanted to go and what I wanted to build.

As you can imagine, creating *your* life of wealth, freedom, and happiness requires more than just vision. Dreams without action remain just dreams.

Before you can become a successful manager, leader, business owner, or entrepreneur, you need to enlist the help of others—your investors, team, suppliers, and clients! It takes *strength of character* to do this, to stay on your journey, and to achieve your wonderful vision.

That brings us to *Chapter 3—On Fire with Purpose, Power, and Passion!*

CHAPTER 3:

On Fire with Purpose, Power, and Passion!

So you can unleash your amazing potential!

"If you realized how powerful your thoughts are,
you would never have a negative thought."
—PEACE PILGRIM, American teacher, spiritual leader, and peace prophet.

"Be still when you have nothing to say;
when genuine passion moves you, say what
you've got to say, and say it hot."
—D. H. LAWRENCE, British poet, novelist, and essayist.

IN CHAPTER 3

- The Three-Point Plan to find your Life's True Purpose—the secret to unstoppable motivation!
- Bored or burned out? Get the Four Rules to Turn Work into Pleasure!
- Discover your Six Power Words—vital energy for courageous, decisive, and authentic actions.
- How to Ignite Red-Hot Passion—inspire your investors, lead your team, and persuade your clients!

THE BIG PICTURE

In Chapter 2, we designed your vision—the business life of your dreams and you living a life you love. If you designed it correctly, it is like the fog clearing, and suddenly you are on a mission to achieve it. Feelings of urgency, excitement, and anticipation come with it!

Now, we shift our attention to the journey—for business success, and indeed life itself, is a journey, and often a difficult one.

To reach that wonderful destination, you must take thousands of timely and correct actions consistently aligned with your vision. To do this in a sustained manner, year after year, requires more than just a mission statement like the ones we see hanging in corporate offices or tucked away in some obscure page on a website.

You need a *compelling mission* in your life—one so inspiring, motivating, and powerful that you are driven to action. With it, you are unstoppable! It must be a mission wherein you know that your actions are things you MUST do and not merely actions you *should* do. Each step must be made toward your vision, and those that follow are so aligned, timely, and correct that there is no other alternative but to act NOW!

A compelling mission is the fuel that gets you out of bed each morning with a spring in your step—not *having* to go to work but *wanting* to work. It absorbs your attention like a sport or hobby. It makes your working day fly by so that, at the end of it, even if you are tired or even exhausted, you say to yourself, "Wow, that was an extraordinary day."

It brings your job to life. It makes work a joy!

With it, you know deep inside at the end of each day that you are one step closer to your vision.

A compelling mission in your life, like *wow* goals, also helps when the journey gets tough. We all know there are days when nothing goes right: You stub your toe within a minute of getting up; your cat vomits at your feet as you head out the door; your clients seem ungrateful and are complaining; or your team seems not to care. A compelling mission makes that tough journey easier and smoother.

A compelling mission is the reason WHY you work, why you want to work, and the very reason *for* your work. Even better, a compelling mission touches you. It brings you alive! It makes your spirit sing. It is like a fire burning in your soul! A compelling mission is your *life's true purpose*—it is what you are here to do. It is that big! So, how does this critical concept integrate with *Wiring Your Brain for Success*?

Remember the Fifth Key: Change must be desired. The greater the desire, the greater the possibility of wiring and rewiring.

The Fifth Key also states that when you are fired up for change, and when motivation and engagement are high, the brain releases chemicals that facilitate the process of brain cell connection and disconnection. Wiring your brain for success is easy.

With a compelling mission, motivation and engagement *are at their highest*, and wiring your brain for success is easiest, quickest, and most enjoyable! So, first up in Chapter 3, I am excited to give you a simple three-point plan so that you can find your life's true purpose—essential for unstoppable self-motivation. Demotivation, boredom, or feeling burned out are very real dangers in business, so I will also share four rules with you to turn work into pleasure.

There are challenges ahead. Other people will place a host of obstacles in front of you, threatening to make your journey more like a long and winding road. You need personal power to overcome such hurdles, and tons of it! So, discover your six power words for business and life that will give you the vital energy to take courageous, decisive, and authentic actions.

Just as importantly, in business relationships you need to inspire your investors, lead your team, and persuade you clients so that they help you—indeed, *propel* you—to succeed at your vision. For this, you need burning passion! So, in Chapter 3, you will also discover how to ignite your red-hot passion! First things first, light a fire in your soul and be unstoppable!

THE THREE-POINT PLAN TO FIND YOUR LIFE'S TRUE PURPOSE

Point 1: Know the best type of motivation

Take a quick look through business bookstores. Motivation is a big subject, and why not? Millions of bosses around the world are trying to motivate their workers to work harder, longer, or more efficiently so that their companies are more profitable.

Wiktionary defines motivation both as willingness of action and an incentive or reason for doing something.[24] Here, the *reason* is critical—it makes sense of your actions. With respect to creating a compelling mission, we are talking about both a *willingness* to take actions (thousands of them) *and* a reason for doing so.

There are two main theories of motivation: extrinsic–intrinsic theory and push–pull theory.[25] Let us look at both in turn.

1. Extrinsic–intrinsic theory:

Extrinsic motivation comes from *external* influences (i.e., from your environment). We are all familiar with the picture of a donkey with a carrot dangling in front of it and a stick wielded behind: reward and punishment. Both objects motivate the donkey to act and move forward; both are extrinsic motivators.

In human endeavors, there are many forms of extrinsic motivation. Money in return for your work or higher grades at school for extra diligence are examples of rewards to encourage the desired behavior, whether for work or study.

Competition is another extrinsic motivator. It encourages the performer to win and beat others. A cheering crowd, the desire

24 "Motivation," Wiktionary, accessed October 8, 2019, https://en.wiktionary.org/wiki/motivation.

25 Adapted from "Motivation," Wikipedia, accessed January 14, 2020, https://en.wikipedia.org/wiki/ Motivation

to win a trophy, and the subsequent adulation of the crowd are powerful extrinsic motivators.

However, there is a problem with extrinsic motivation, both for businesses and individuals—they often do not work or, worse, can negatively affect motivation and performance. For example, setting targets in a team setting (with rewards "dangled") may set off counterproductive competition within the team or resentment if the targets are not reached. For an individual, failing to meet a goal can be disheartening or demotivating. Extrinsic motivators can give us more of what we do *not* want: demotivation!

While extrinsic motivators are still important, in his book *Drive—The Surprising Truth About What Motivates Us*, Dan Pink lists the key downsides of extrinsic motivation: it can extinguish intrinsic motivation (see below); diminish performance; crush creativity; crowd out good behavior; encourage shortcuts, cheating, and unethical actions; become addictive; and foster short-term thinking.[26]

Conversely, intrinsic motivation comes from *within* you. If you are intrinsically motivated, you have the self-desire to act. Examples of intrinsic motivation are seeking new challenges, gaining knowledge, helping others, or working when work is a pleasure. Play, curiosity, humor, and gratitude are other examples of behavior that can be intrinsically motivated.

Pink lists three main components of intrinsic motivation (what he calls Type I behavior), which includes autonomy or being self-directed, mastery or the devotion to become better at something that matters, and (what do you know?) *purpose*—the quest for excellence linked to a larger or higher outcome.

---

26 Daniel Pink, *Drive: The Surprising Truth About What Motivates Us* (Edinburgh: Canongate, 2011).

2. Push–pull theory:

Push motivations are present when you push yourself or are pushed toward a goal, or when you are pushed to achieve something. This includes the pressure to gain financial independence; being told to rest, relax, or get fit; or having to attend a social function because of your obligations.

The big disadvantage of push motivations is that obstacles can easily discourage the path toward achievement. Push motivations can also be a negative force. We can all relate to the pushy salesperson that puts us off making a purchase by trying to *push* us into a decision.

Pull motivation is the opposite and is much stronger: It is the desire to achieve an outcome so badly that it seems the goal is pulling us toward it. For example, a holiday to an attractive destination seems to be pulling us toward a particular destination—a tropical island escape could lure us in with pictures of palm-fringed beaches, coral reefs, action adventures, nightclubs, or simply relaxation and rejuvenation. We may feel *pulled* into working out in a gym by thoughts of a fitter appearance or *pulled* to attend a social function because you enjoy the company of the people who will be there.

Pull motivation can be a positive force, overcoming the negative influences of a push motivation. To use our example of the tropical holiday, the colorful picture of the palm-fringed beach might pull us into making a buying decision despite the push of an over-enthusiastic travel agent.

To return to the subject of defining a compelling mission for our business lives, which types of motivation are likely to be more inspiring, sustained, and, motivating: extrinsic or intrinsic? Push or pull? It is *intrinsic* or *pull* motivations that give us the energy to drive the actions that get the results we want—steps on the journey to our vision.

At one point in my business life, prior to becoming a sole owner, I was completely burned out. Although the technical aspects of my job still fascinated me, I was working long hours for little pay, and I was carrying a lot of business and personal debt. For their own reasons, my business partners did not want to be working within the business. The partnership was dysfunctional, and the tension was unbearable.

One day, on duty and working alone in the office late into the weekend, something snapped inside me. I literally screamed in frustration. It did not make me feel any better. I screamed again. The walls were silent witness to my frustrations. I screamed a third time, even louder. The empty building echoed with my howl of mental pain.

My frustration turned to rage. I got up from my desk and paced from room to room, white with anger, slamming one door after another. Eventually, I kicked a door, partially caving it in.

None of my actions did any good. I just had to get on with my work and life. I knew I had experienced the symptoms of acute and chronic burnout that afternoon. I felt empty, lonely, and exhausted.

Looking back with the wisdom of hindsight, I was *externally* motivated. I had to work to pay off debt; I had to work to keep the business partnership running; I had to work to support my young family. However, all those external pressures had taken the pleasure out of work. I was *pushed* into working by my bank, driven by its unspoken threats of holding everything I owned as security. I was *pushed* by my business partners' expectations and by the responsibilities I had as a husband and father.

Expectations and responsibilities are often unavoidable, but my anguish at the time is a graphic illustration of the debilitating effects of a lack of *intrinsic* motivation and the complete absence of a compelling mission *pulling* me toward a magnificent vision. I wish I had known what I know now.

How many countless thousands of businesspeople have been in my position? Their details will be different, but the core reasons for their plight will be similar.

Do you have a compelling mission intrinsically motivating you, pulling you forward to take actions, or are you consistently demotivated or, as I was, burned out? I now understand that my problem was ignorance. I was not at fault for feeling as I did. It was caused by a complete absence of knowledge about motivation and what to do when motivation is fading or gone. In fact, the factors causing my distress were completely understandable.

Here is an important point: Financial, business, and personal pressures need not cause demotivation, exhaustion, or burnout *if* you are driven by a compelling mission. Little did I know that motivation can be reignited, and you can find a compelling mission in your life—indeed, your *Life's True Purpose*—even while performing the same job! New desire can ignite a flame within your soul, burning with new intensity and energy, driving your actions, and motivating those around you.[27]

Think about your *current* business life. In what percentage of your working week do you feel intrinsically or pull motivated?

In your dream business life (as defined by your vision), in what percentage of your working week would you *like* to feel intrinsically or pull motivated?

---

27 If you are interested in maintaining and improving intrinsic motivation as an individual and within a business, Dan Pink's book is an excellent starting point, as he has developed toolkits to help you.

Reflect on the gap between your two answers.

Time for Step 12 in the online course!
It can be quite revealing!

As you can imagine from my story, there was a *massive* gap between my motivation at the time—predominantly the extrinsic, push type—and the ideal of intrinsic, pull motivation.

Points 2 and 3 detail how I closed the gap.

Point 2: Know what motivates YOU

Defining your *compelling* mission is a central theme of this book and the companion online course. It is critical because key outcomes—success, freedom, wealth, and happiness—flow from the discovery or fine-tuning of, or refocusing on, YOUR mission.

Your compelling mission operates like an internal guidance system *pulling* you toward your personal vision. Without one, you risk floating aimlessly through life, never really reaching your amazing potential. Understanding exactly what motivates you is critical for sustained energy, productivity, and progression.

When it comes to understanding personal motivations, people often fall into three categories: the lucky ones, the metal detectors, and the lost dogs.

The "lucky ones" go through life knowing what they were put on this planet to do—the ones who seem so lucky and full of confidence and for whom life seems served on a plate. They are people who appear to have integrated the perfect job or business with the perfect life.

Many of the lucky ones have reached this point through sustained and intentional effort, and their "luck" is really the reward for hard work. However, for some lucky ones, this is not the case. Their outer confidence is not matched by inner certainty. They could be like me many years ago, at risk of burning out.

The "metal detectors" are people with a remarkable ability to locate and pursue new and shiny objects on an astoundingly regular basis. They are always searching for that undefined something that continues to elude them as they swap one job for another. They are tempted by an ever-increasing number of vocational options available to them, never committing to one course of action or finding that special something that is truly "theirs."

Metal detectors throw themselves with stunning intensity into their jobs, full of contagious enthusiasm, only to lose interest and quickly move on to the next grand opportunity.

Finally, the "lost dogs" seem to wander through life never experiencing deep happiness. Their job is just a job, a means to an end, and the end is the money to spend on the rest of their lives. It seems a shame that they never quite reach that "time" when they live their dream. What a waste!

Can you relate to the lucky ones, the metal detectors, or the lost dogs? Perhaps there is someone you know who fits into one of these categories.

We spend so much of our lives at work, and sometimes its technical aspects can obscure our true purpose. The architect designs buildings; the football coach coaches a team; the accountant prepares financial statements; the software developer writes code; and the brain surgeon operates on brains. No matter your job, the technical aspects have repetition to them. Whatever you do, you do it time and again until you become exceptionally good at it—a master! Nevertheless, you keep doing it—that is your job.

Lost in the repetition, is it clear *why* you do what you do? What are your deeper motives? Do the technical aspects of your job still provide *deep inner fulfillment*? Have you thought about the *reasons* why you selected your vocation? Did it just seem like a good idea at the time, or were there deeper motives behind your decision?

It was my mother who suggested my first career. Her suggestion had never occurred to me, but when she recommended it, it made perfect sense.

Looking back, I can clearly identify the three motivations that guided me to follow her advice.

I thought my chosen path would be a career with plenty of variety and analysis, so I was motivated by the newness and discovery that was an integral part of the job. I knew that although it was not among the highest paid professions, I would earn a good income, so I was also motivated by money. Finally, this profession seemed to be held in high esteem within the community, so I was motivated to some degree by the thought of recognition and prestige.

My choice seemed like a good idea at the time, but many years later, this combination did not generate sustained and successful motivation. My burnout was a graphic testament of the need for clarity about the reasons WHY you work.

Reflect on the gap between the degree of motivation you currently feel and the intrinsic or "pull" motivation in the business life of your dreams.

Are you inspired and sustained every day, month, and year? Do you know your *life's true purpose*? Will it lead you to red-hot passion and amazing business success? Will it give you the freedom to do what you want to do? Will it make you wealthy? Will it bring you deep and abiding happiness?

These are important questions, as your compelling mission will give you the energy, intensity, and drive required for the journey ahead. What motivates you?

Let us find out. There are many motivations in the field of human endeavor. These include:

1. Desire
2. Power, winning, influencing, controlling, competitiveness, and dominance
3. Discovery and newness
4. Excellence and achievement
5. Acquiring knowledge and being smarter
6. Self-improvement
7. Self-achievement
8. Caring for others and nurturing
9. Sex, beauty, love, and romance
10. Prestige and owning the best or most expensive objects ("I am better than you")
11. Saving time
12. Having a great time, entertainment, and fun
13. Service to others
14. Money and wealth
15. Fame
16. Praise and recognition
17. Perfection
18. Building empires (small, medium, and large businesses)
19. Family
20. Wanting to belong.

This list can seem daunting, as so many are tempting! To know what motivates you, use the following four rules.

THE FOUR RULES TO TURN WORK INTO PLEASURE

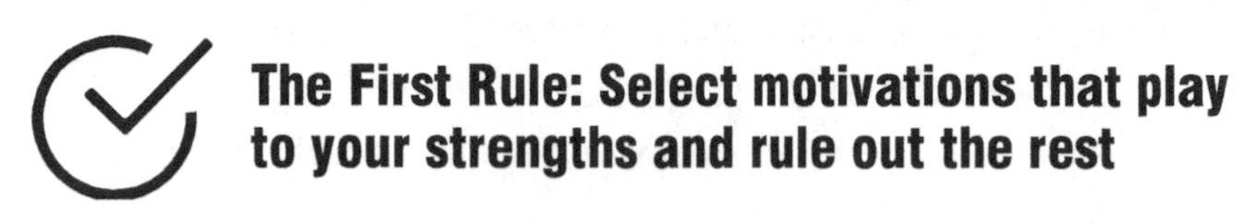

The First Rule: Select motivations that play to your strengths and rule out the rest

Remember the Fourth Key: Practice makes perfect.

Elite performers "play to their strengths." From an early age, when their natural skills were becoming obvious, they learned to strengthen and increase the brain cell networks responsible for their abilities *through practice* until they achieved mastery in their endeavor.

If you want to be an elite performer and achieve mastery in your area of business, then select motivations that utilize your strengths. Reject those in an area of weakness, even if it is a tempting choice.

You are then focusing on the positive thought processes that make you the unique and amazing individual you are—indeed, converting your strengths to superpowers.

Top athletes, actors, and musicians did not achieve their mastery by focusing on their weaknesses. Imagine Usain Bolt trying to play championship tennis or Serena Williams trying out for the 100-meter sprint in the Olympics. They would not have become the champions they are. They concentrated on their STRENGTHS.

It is no different in business. Be motivated by your strengths.

To be a real champion in your business life, turn good into excellent, and then turn excellent into amazing!

The Second Rule: Select motivations that have deep meaning and rule out the rest

When we are working on something we care strongly about, which generates positive and passionate feelings within us and has deep and heartfelt meanings for us, we are likely to give it our full attention.

> *Remember the Ninth Key: It is as easy to cause negative rewiring in your brain as it is to wire in positive changes, and the Eleventh Key: Positive emotions easily, instantly, and positively wire your brain.*

It is imperative, then, to have deeply and meaningful motivations. Without meaningful direction, our brains can flip-flop between negative and positive rewiring without making any progress. We can easily become sidetracked by all manner of distractions. The risk is multiple career changes without progress, little progress in one career, or little focus on what should be important.

What made people like Nelson Mandela and Mother Theresa so special and, in their own way, so successful? They CARED deeply. Feel moved by your choices! Do what you care about.

The Third Rule: Do what you love and rule out the rest

Millions of people are trapped in jobs or vocations that give them no pleasure. Is it any wonder that their path to success is blocked by an absence of sustained and successful motivation, let alone a compelling mission? They may have dreams of a wonderful future, but their vehicle lacks the fuel to make the journey, let alone a boost from the high-octane version of a compelling mission.

> *Remember the Fifth Key: Change must be desired. The greater the desire, the greater the possibility of positive wiring and rewiring.*

If you want to create your dream business, then select *and look forward to* building a career around something that gives you an amazing amount of pleasure, continuing your journey with *desire*. What could be better?

We all know that when we are absorbed with something that gives us pleasure—be it work or play—time flies. At work, we seem to achieve so much more. Our progress is greatest, and motivation is high. Conversely, when we have little or no interest, and our work gives us no pleasure or even feelings of dislike or hate, time drags, progress is slow, and motivation is low.

Think how passionate people sound when they talk about the special something that gives them enormous pleasure. In comparison, we have all met salespeople who generate false passion to make sales because their own motivation is lacking. They attempt to *push* you into making a purchase rather than allowing you to be *pulled* into the decision.

People like Richard Branson and Oprah Winfrey are so inspiring because it is obvious they gain so much pleasure from what they do. They LOVE what they do!

Do what YOU love; listen to your inner voice—go where it takes you! It sounds great! It *is* great!

The Fourth Rule: Select only two or three motivations

This can be difficult. If you have not already done so, narrow the list to two or, at most, three motivations. Why is this important?

> *Remember the Second Key: In your mind, what you focus on grows bigger and bigger.*

What progress would you make if you focused on all 20 motivations on the list? None! The positive effect of focus would be too diluted. It would be like trying to be good at 20 different sports or career pursuits at one time.

Conversely, what positive progress would you make if you focused on just one or two? Massive! How easy would that make your business life? What time could you save?

Look at some famous business leaders: Bill Gates, Warren Buffet, and the late Steve Jobs—all are or were highly FOCUSED. Develop a laser-like focus!

How did I make the change from being burned out to having a compelling mission *without changing jobs*? How did I rekindle the enthusiasm I felt as a student or new graduate? How did my work become pure joy?

After my burnout, I drifted for many years in my job. It was a daily struggle to get motivated, but the variety was still so interesting and the income reasonable, so I ploughed on, swinging between heartfelt motivation and demotivation. Two of my motivations—newness/discovery and money—were still relevant to me. The other motivation—recognition/prestige—now held no attraction.

Even when I was burned out, I knew there was a hidden, powerful motivation I had not internally identified or recognized, let alone verbalized. Something must have been keeping me going. The feeling nagged at the back of my brain, waiting to surface and say, "Hey, pay attention to this inner voice that is telling you something important."

In a weird way, I was aware of this motivation all along, yet I ignored it. I had to go back to my roots, back a whole generation, to unearth it and work with it, shaping it into something that is now the driving force within me. I had to go back to my father and what his family had taught me as a child.

Dad was one of three brothers, and they all worked in banks their entire working lives. Dad's two brothers worked their way up into top executive positions while he was content to stay behind the counter, serving customers until he retired. Irrespective of their relative positions, they were bound by a similar ethic—one of service to others.

Slowly, this theme of service began to assume greater importance in my mind and working day. Every time I did something good for someone else, they seemed to feel "the good," and so did I. I felt a little more energized and motivated and a little less burned out.

It did not seem to matter if the person was a client, one of my team, a supplier, or an acquaintance. The tiny win-win made my day a whole lot better.

It did not seem to matter if I had done an amazing job for someone and they were very grateful for my work or if it was a simple act of kindness, such as comforting someone when they felt low—the feeling was the same.

It also did not matter whether a large amount of money or *no* money was at stake because goodwill between humans was generated, and I felt great.

This feeling generated other wonderful feelings: confidence, generosity, gratitude, optimism, and happiness. Like a snowball gathering mass as it rolls down a hill, my whole outlook at work and in my personal life slowly changed.

Eventually, each day was filled with my acts of kindness or goodness. Each time someone thanked me, I replied with complete sincerity: "That *really is* my pleasure"—and it still is.

As you read this, it is reasonable to ask: "Is this path not totally self-sacrificing? What about one's own needs and wants?"

Each day, my work brought me new joy. Each day, I found new ways to create acts of kindness and goodness. In fact, the opportunities to do so were limitless. So, what about my own needs and wants? I am far richer than I ever was, and in every way. I have never enjoyed life more! I had discovered my hidden motivation.

What about my original motivations?

Money was still important to me, but my attitude to money had changed completely. I now viewed money simply as a measure of how much value and goodwill I created for others. It was a major shift in my belief systems.

Newness and discovery were still important because the technical aspects of my work still fascinated me. I forgot about recognition and prestige.

So, these are my three motivations: service to others, money, and newness and discovery.

They comply with the Four Rules to Turn Work into Pleasure: Serving others is one of my strengths; I care deeply about them; I love what I do; and I have a laser-like focus on the task at hand. What a transformation—from burned out to discovering the compelling mission of my life!

**It is your turn! What motivates you?
Head over to Step 13 in the online course
to find out! It is *so* helpful!**

Now, let us return to the last point in the Three-Point Plan to find your Life's True Purpose.

Point 3: Craft your personal mission statement—it is your Life's True Purpose

Knowing your top motivations is a key to success. However, it is difficult to keep them uppermost in your mind as you go about your daily work. As I have said, we all have bad days when nothing seems to go right. How do we turn what threatens to be a bad day into a good one and make a good one even better, creating one amazing experience after another?

How do you turn bad into amazing in seconds? How do you maintain sustained, successful intrinsic or pull motivation? How do you turn motivation into a compelling mission? How do you transform a compelling mission into your *Life's True Purpose*?

You need a simple phrase that:

1. Combines your top two or three motivations;
2. Obeys the Four Rules to Turn Work into Pleasure;
3. Is easy to remember;
4. Is anchored in rational, proven science using The 12 Keys;
5. Through repetition with daily use will achieve further positive wiring and rewiring;

6. Makes sense and feels right to you;
7. Creates a feeling of intrinsic, pull motivation;
8. Stands the test of time, generating sustained successful motivation;
9. Can readily turn a bad day into an amazing one; and
10. Is one phrase or one sentence—the shorter, the better!

This is a personal mission statement. It is a simple statement of your *Life's True Purpose*. Its true power is its ability to motivate you and focus your attention on what is important every moment of the day. It is the reason WHY you work. It turbocharges your brain for success.

With it, you are unstoppable!

How did I discover my life's true purpose? I realized that I had to incorporate my three top motivations:

1. I decided that service to others was my top motivation.
2. One of my key strengths is creativity, and creating something (no matter how big or small) has an aspect of discovery and newness. Thus, incorporating the word creativity combined both a personal strength and my motivation of newness and discovery.
3. Money is a measure of value. Therefore, rather than focusing my attention on money or wealth, I used the word value.

Combining my top motivation (service to others) with the concept of value, I coined the phrase value to others. I then combined the three concepts together—creating value for others. This included my three top motivations and a key strength in just four words—I liked it! (It sounds easy now, but it took many attempts and revisions.)

The next step was to quantify and qualify the concept of value.

I knew that the greater the service (or value) I performed for others (whether money was exchanged or not), the better I felt. Just how much value is required for sustained and successful motivation? Well, *massive* amounts of value were likely to generate massive goodwill and mutual benefit, so I incorporated the word "massive" into my statement—creating massive value for others.

Finally, I knew that the last thing I wanted to do was create massive amounts of substandard or average value. One of my core business values is the concept of quality; whatever I do, I instinctively try to produce the highest quality result. Just how much quality is required for intrinsic, pull motivation? Imagine a standard of quality that amazes people! That is the standard I strive for. Putting it all together, this is my personal mission statement, my *Life's True Purpose*: to create massive and amazing value for others.

It did not matter whether I was working in my business or, indeed, writing this book and the companion online course. This simple phrase described my compelling mission perfectly.

It gives me joy. It produces sustained, successful, intrinsic, pull motivation. It incorporates my three top motivations and obeys the Four Rules to Turn Work into Pleasure. It is easy to remember, makes sense to me, and can turn a bad day into an amazing one. It is WHY I work, the one big reason why I am here on earth. Yes, it is *that* big!

Some people (and I am one of them) *love* what they do and do what they *love*. Love is, of course, the strongest of all emotions, and if it is an integral part of your compelling mission, then it will give you the strongest of motivations. You will move mountains and overcome all manner of obstacles to achieve your vision.

Love does not have to be part of your mission, but if it is, it is absolutely wonderful. I *love* creating massive and amazing value for others!

What is your Life's True Purpose? Use Step 14 in the online course to discover it!

Step 14 is critical to your success.

This is your hardest work, and only you can do it. Again, listen to your heart, spirit, and soul. Listen to your gut instinct. Think deeply and be patient.

You may have to revise this multiple times, taking many drafts to reveal your *Life's True Purpose*. Once you have a robust version, repeat it to yourself often.

Use the Fourth Key to wire your brain for success: repetition! Slowly embed it in your psyche.

Your vision will be a gigantic, exciting network of millions of brain cell connections in your mind—indeed, a detailed instruction manual or *model* of your future—so it is with your compelling mission statement.

If you design it correctly, it will also be a complex network of millions of strong and enduring brain cell connections. To return to the computer analogy, this model is like a software program of *logic* describing the underlying reasons for *why* you work; it is very considered, intensely focused, and designed to be powerfully motivating and make you unstoppable!

> *It is the Sixth Key in action! Remembering models guides positive wiring and rewiring.*

If you further strengthen your compelling mission using all 12 Keys, your brain will be *hardwired* with a positive model to guide your beliefs, attitudes, and behaviors. Indeed, it will be a model to *drive* your future actions—and actions get results!

Like mine, your compelling mission may change as the years unfold, but knowing it makes all the difference. Find it and make a gold mine from it!

THE CHALLENGE AHEAD

Your personal vision is the destination for *your* journey. Your compelling mission is *your* inner drive to make it.

However, the road ahead is not easy or straight. To make the journey, we need to enlist the help of many *other* people along the way, including our life partners, business partners, investors, teams, suppliers, and clients. It is highly likely that many people are just not interested in your journey, and reasonably enough—they have *own* journeys to make. Who are we to impose ours on them? The personalities of key people such as your partner, team, and clients will be a key issue for you.

It is highly likely that they will, unlike you, be unaware that they can wire their brains for success using the knowledge of neuroplasticity. As they go about their daily tasks, they will flip-flop between positive and negative thought patterns, affecting what they do and how they see their world. Their thought patterns, which are different to yours, will form their beliefs, behaviors, and actions, as well as the results they get on their journey.

If indifference to your journey were the only obstacle, it would still be straightforward. However, you are likely to meet every possible negative human emotion along your journey as well as, thankfully, every positive one.

While you will meet people who will love you for what you do, you will also meet people who hate you, no matter how nice you are. They will hate you for your niceness, and they will hate you for your success. You will meet both generous and mean-spirited people, those who wish you every success you so richly deserve and those who are so jealous that they will try to sabotage any progress you make. You will encounter kind souls and bullies, saints and criminals, those who cheer for you, and those who could not care less, despite the fact that you poured your heart and soul into what you did for them.

Inevitably, there will also be differences of opinion. You cannot win all arguments, so how do you know which fights to pursue and which to let go? When is it wise to push ahead and when is it appropriate to walk away?

Other people can accelerate or slow our progress, block us, set us back, or even send us in the wrong direction. They can be so frustrating! How do we navigate our way forward?

What you need is personal power—the *fuel* within to keep going despite all the negatives other people place in front of you. You may not realize it, but you have that power within you now, perhaps not yet formally identified or verbalized.

It is your *values*!

SIX POWER WORDS FOR BUSINESS AND FOR LIFE: YOUR VALUES

This is a key point: Values are *what* we stand for. They give you the vital energy for decisive, courageous, and authentic decisions. Your values are your principles and standards—your judgment of *what* is important in life.

They influence every aspect of our lives, including our sense of *what* is right and wrong, our actions and answers to others, and our commitments to goals at work and in our personal lives. They set the limits or boundaries for the hundreds of decisions we make every minute of every day—what to do and what not to do; when to push forward and hold back; and when to say yes, no, or nothing at all.

They help us fully understand the reasons behind our decisions.

When your values have not been identified, lack clarity, or are suppressed, you lack personal power. Like other people, you flip-flop between positive and negative thought processes, and your vision is submerged, along with everyone else's dreams and aspirations. You remain mired in mediocrity.

A clear set of values gives you five big advantages:

1. Values inject and magnify personal power. Once armed with a magnificent vision, a compelling mission, *and* your values, you have the personal power to stand up for your beliefs. If one of your values is violated (in your personal or business life), it is easier to make a stand. By doing so, you keep heading in the right direction, toward your vision. You are in control of your life when you are clear about your personal values. Values guide and *model* our beliefs, behaviors, and actions. *Remember the Sixth Key: Models guide positive wiring.*
2. Values are like signposts, setting direction for the journey we are taking. Having clarity about our values makes it easier to give direction to others. A businessperson with a clear set of values might say, "Look, this is a key issue here. What is at stake is one of our agreed values. This is the way we do it around here." *Values point the way.*

3. Values motivate. If one of your values has been threatened or violated, and you stand up and voice your concerns, you are moved by your values. It is a great example of intrinsic motivation at work.
4. Values keep you focused on what is important.
5. Values encourage a sense of good, ethical, moral behavior in you and others.

It is essential that the journey to your magnificent vision is grounded in the highest sense of ethics and morals. If you bend the rules and fall from grace or lose the trust of key people, it is hard to regain momentum. Trust is like a Ming vase—once broken or even cracked, it is never the same, and its value diminishes.

In business, values are an integral part of success. Common values shared by a team encourage and enable fellow members to act independently and make their own decisions within the framework of the business structure and strategy. Teams are far more effective when shared workplace values are clear, understood, *and implemented* by all.

I will never forget one occasion. I was holding a team meeting, and everyone was there. The agenda had been distributed early, so everyone had a heads-up.

One of the items for discussion was an issue I felt strongly about. I was going to suggest a change in policy that would alter the advice the whole team gave to our clients. I needed most of the team on board.

As the owner of the business, I was chairing the meeting, so I launched into the topic, talking positively and passionately about the subject. When I had finished my short presentation, I stopped and looked at my team for feedback, hoping to reach broad agreement.

One of our team leaders looked me in the eye and said, "Well, Rod, you obviously have not read the recent journal findings, have you? The latest research directly contradicts what you are proposing."

"Err..." I started searching for words, caught off guard, looking around the room for support.

"Well, it's in your in tray," he continued. He held up *his* copy of the journal for all to see as proof of his assertions.

"Ahhh..." I was floundering.

I felt heat spread across my face, blushing bright red in front of my team. Sweat broke out on my forehead. I could not stop my emotions, on display for all to see. It was a total embarrassment.

"Well, let's return to the item at a later date," I said lamely, and moved on, unable to verify what he was saying.

He was right. I had not read the journal article; ironically, however, my "team" leader was quoting from only one study, and one study does not constitute solid scientific proof. In addition, his article was only one aspect of a broader subject I still believed in.

This was not the point. He had intentionally humiliated me in front of my whole team. Further, he had had ample opportunity to discuss the issue prior to the meeting and resolve it without conflict.

In fact, the real issue ran much deeper. In blatantly disregarding my feelings, he disrespected the business model we had taken so long to build. Empathy was one of our agreed team values, underpinning everything we stood for.

If trust is like a Ming vase, my trust for him was smashed into tiny pieces.

Given that I had identified empathy as a key value, I knew he likely had no future in my business. I may have lost the fight, but I won the battle. Shortly after the incident, he left my business—probably realizing he was not a good fit!

How do you identify your values? Here is a list of suggestions. Please add other values that are important to you.

Compassion	Freedom	Joy	Life
Peacefulness	Hope	Family	Health
Fidelity	Truth	Trust	Enthusiasm
High Energy	Strength	Courage	Gratitude
Positivity	Creativity	Power	Friendliness
Relaxation	Care	Honesty	Quality
Application	Openness	Sincerity	Well-being
Confidence	Humor	Love	Happiness
Excellence	Excitement	Pride	Contentment

As with motivations, there are too many choices to remember or use in everyday life. The list must be narrowed down. Six or seven is the ideal number.

As with your vision, you need one set of values for business and another set for your personal life. Some values may be common to both "lives," but others will necessarily be different.

It is important that your business values are consistent with your vision and compelling mission. Each one should complement the other in perfect synergy. Each set of values is your Six Power Words—vital energy for courageous, decisive, and authentic actions!

Just like the situation when I was embarrassed by my team leader, every time an issue arises that involves one of these six words, your brain should change a gear and ask, "Is this issue consistent with my values? Do I need to voice my concerns? Do I need to make a stand? Do I need to take action?"[28]

---

28 At the time of my confrontation with my team leader, I failed to act, even though one of my values had been violated. I had not yet mastered my thoughts and emotions, and my fear of conflict was still very disabling. See Chapter 4.

Once you have selected your values, expand on each one. Think about what they mean to you and how you apply them in your business or personal lives.

As an example, here are two of my personal values and the effect each one has on my life:

The first is *creativity*. As a business owner and entrepreneur, I can create products and services that provide happiness, income, and wealth to others. As a creative person, I can bring happiness and laughter to my family, friends, and acquaintances. I am most joyful when I am being creative.

My second value is *peacefulness*. Inner peace, outer calmness, and harmony with my environment enable me to focus on what is important at each moment with a clarity of thought. It frees me from past self-imposed limitations, hurt, inhibitions, anxieties, and fears. It enables me to be creative, playful, and happy. By being peaceful, I provide one small example for others to follow.

What are your values? Get clarity with the valuable Step 15 in the online course!

Just as your vision and compelling mission are networks of brain cells in your mind, so too is each individual value. If you choose them wisely—so they sit well with you—and define them clearly, your values will also be complex networks of thousands of strong and enduring brain cell connections.

To return to the computer analogy, each one will be like a software program, this time of your *standards*. Just like your mission, they will be very considered, powerfully motivating, and intensely focused.

Again, it is that Sixth Key in action: Remembering models guides positive wiring and rewiring!

As with your mission, if you further strengthen your values using all 12 Keys, your brain will become *hardwired* with positive models to guide your beliefs, attitudes, and behaviors, with each value *powering* your future actions. Once again, actions get results!

HOW TO IGNITE RED-HOT PASSION

Have you ever asked this question: Other people are successful, but why not me? One factor in their success is their red-hot passion. So, how do *you* find that red-hot passion? This is my formula:

Vision + Mission + Values + The 12 Keys = Red-Hot Passion.

Use it to find your voice!

It is through your voice that you engage and inspire others. Give them direction and leadership with your vision; motivate and persuade them by sharing your compelling mission; and show confidence and independence with your strong sense of values. Point the way!

Get this process right, and you will tremble with raw emotion. A powerful feeling will rise from the depths of your heart and soul and will lift your spirit. This is the measure of your care. It is red-hot passion! Ignition!

In that trembling moment, consider this question: Just how wonderful could your future be?

Patterns of reason and positive thought processes are now wired in your brain—easy to remember and so incredibly simple. Yet, for some, without the knowledge to change the way they think, feel, and see the world, life is *so* difficult. They flip-flop from positive to negative thought patterns and remain mired in mediocrity or drown in negativity and failure.

Your vision answers the "where" and "when," your mission answers the "why," your values answer the "what," and The 12 Keys tell you "how!" Now you have a step-by-step formula to create your unique brand of red-hot passion and, with it, the power and energy to create the business life of your dreams—and a life you love!

You are on your way!

KEEP IT SIMPLE

- A compelling mission is so powerful that it compels you to act.
- Use this three-point plan to create one:
  1. Know that intrinsic, pull motivation is essential; it is self-sustaining and successful and comes from within.
  2. Select your top motivations. Make sure your motivations give you real pleasure, play to your strengths, and have deep meaning for *you*. Limit your list to two or three motivations.
  3. Create your compelling mission statement using your top motivations. This is an easy-to-remember, short phrase or sentence that creates intrinsic, pull motivation. It makes sense to you and has the power to turn a bad day into an amazing one. It is the reason why you do what you do; it is your *Life's True Purpose*! With it, you are unstoppable!

- Values are our principles and standards, our judgment of what is important in life and what we stand for. They are both a source of personal power and signposts pointing the way to your vision. Identify and expand on your top six values, one set for both your business and personal lives.
- Combine vision, mission, values, and all 12 Keys to find your voice. It is through your voice that you engage others on the journey to your vision. This is the source of red-hot passion.
- Use your passion to inspire your investors, lead your team, and persuade your clients!

TRANSFORM YOUR COMMUNITY

Communities around the world are dying due to lack of vision, mission, and values. Often, it is simply a lack of correctly applied knowledge that is the difference between success and failure, growth and stagnation, and business life and death.

We are drowning in a flood of information but suffering from a drought of wisdom. Who but our businesspeople will make a real difference at the coalface where people work hard every day so they can have a life?

What if every manager, leader, owner, or entrepreneur had a magnificent vision for their business, a compelling mission forged in clearly articulated reasons why, an inspiring set of values, the personal power to stand for what they believe in, and red-hot passion that enables a courageous voice? Would it make a difference to you, your community, or your country? YES!

THIS IS ABOUT YOU

If you are aspiring to be a successful businessperson or already are one, who but you are going to make that difference? If everyone waits for the next person to do it, we really will have business death by a thousand cuts. This starts with you!

If you feel the heavy weight of responsibility suddenly land on your shoulders and begin to buckle under the weight, do not! This journey to your wonderful vision—supported by your compelling mission, the personal power of your values, and the intensity of your red-hot passion—is sensational. It is amazing! Fasten your seat belts, turbocharge your brain, and lead the way.

Here is the best bit: You win!

NEXT:

If the journey to a magnificent future was easy, everyone would be doing it. It can be a long and twisting road, especially at the start. Like other endeavors that challenge us, the journey itself can be rewarding. However, progress is often two steps forward and one step back, or four steps back before six steps forward.

Those backward steps can be overwhelming, devastating, or even defeating. Old insecurities, doubts, self-imposed limitations, anxieties, and fears can rise, time and again, seemingly drowning us at the very moment of success. How do you crush them into fading memories, no longer blocking your progress?

Perhaps you glimpse your greatness—how do you unleash it? How do you tap your amazing potential? Such questions bring us to *Chapter 4—How to Master Your Thoughts and Emotions.*

CHAPTER 4:

How to Master Your Thoughts and Emotions

So you can live a life you love!

"There is only one corner of the universe you can be certain of improving, and that's your own self."
—Aldous Huxley, English novelist and critic, 1894–1963.

"The true key is a trust in self, for when I trust myself, I fear no one else. I took control of my life, just as anyone can. I want everyone to see it's in the palm of your hand. The past is gone, the future yet unborn, but right here and now is where it all goes on."
—"The Update," The Beastie Boys, American rap-rock/hip-hop band.

IN CHAPTER 4

- The X Factor in Business Success: What it is, how to get it, and how to maximize its effect. Do not miss this—it is 50% of your results!
- What the gurus should teach: The Three Stages of Personal Transformation.
- Discover the "Software" to Program Your Mind for ________! (You insert the outcomes you desire.)
- Conquer pressure and feelings of being overwhelmed with my Emotional Mastery Formula.

THE BIG PICTURE

How *do* we turbocharge our brain for amazing business success? How do we learn to think, feel, and see our world like people who have already achieved freedom, wealth, and happiness? How do we take the actions they take to get their great results? Business success critically depends more on your emotional intelligence (your *EQ)*—how you control your emotions and manage the emotions of others—than your IQ.

In the Hollywood science fiction blockbuster *Star Wars*, the little Jedi Master Yoda teaches a young Luke Skywalker how to harness the Force in the battle of good over evil by controlling his mental state. With intensive training, Luke begins to regulate his emotions, overcoming his doubts and fears, and, with mastery, he can powerfully influence the minds and emotions of others.

While moving inanimate objects with the power of his mind is the stuff of science fiction, in Chapter 4 you will, like Luke, learn how to powerfully control your own emotions and positively influence the emotions of others for good and great outcomes! We will identify your emotional strengths and, using The 12 Keys, learn how to *turbocharge* them! We will identify your emotional weaknesses and learn how to master them.

> ***You will have an exciting opportunity to draw on the Eighth Key: Mental rehearsal, meditation, and mindfulness rewire your brain.***

Like Luke, we will explore the ancient practice of meditation, but with a very modern take. You will learn how to:

- Draw upon powerful, positive emotions when you need them most in the busy and stressful modern business world; and
- Influence the emotions of others—notably your team, suppliers, and customers or clients.

Using another Hollywood sci-fi blockbuster analogy, *The Matrix*'s protagonist, Neo, receives intensive training from his mentor, Morpheus. With Neo's nervous system plugged into the matrix, Morpheus downloads thought patterns directly into Neo's brain. Extraordinary karate moves and advanced weapons training take seconds to acquire. Want to levitate? No problem. Dodge bullets? Easy!

If only we could train our business minds so readily! Want to maintain team enthusiasm in the face of persistently bad results? Download the Optimism Thought Program and create great team performance! Lost your temper recently? Get the Self-Control Thought File and maintain your cool in the face of extreme provocation.

With the explosion of knowledge in areas such as neuroplasticity, positive psychology, and meditation, fiction is *much* closer than we think. So, *get wired for success*!

Luke, focus your mind! Neo, fasten your safety harness and plug yourself in! This will be exciting!

THE X FACTOR IN BUSINESS SUCCESS: EMOTIONAL INTELLIGENCE

With intensive training and subsequent experience, most of us become experts in one or more fields. You can become a software designer, a dentist, or a plumber—insert your vocation here!

You use your IQ to learn and master your chosen profession or professions. However, we all know that your ultimate business success depends little on your grades at school and college. Even high IQs and graduating with honors have little to do with *business* success. We all know fellow students who struggled to get through a course but became brilliant success stories after graduation, whereas others were first-class students but hopeless businesspeople once they ventured into the real world, sometimes even failing.

If you think your business success is tightly linked to your technical skills or even your IQ, think again. Daniel Goleman's research for his book *Working with Emotional Intelligence* highlighted that *EQ* was *twice* as important as technical expertise or IQ for business success. EQ accounted for 67% of the abilities deemed necessary for superior performance in leaders.[29]

Essentially, your EQ is the ability to recognize your own and other people's emotions, identify different emotions correctly, and use this information to guide thinking and behavior in yourself and

29 Daniel Goleman, *Working with Emotional Intelligence* (New York: Bantam Books, 1998).

others.[30] It is critical to realize the importance of this information, as *your EQ guides your thinking*. In turn, your thinking governs your beliefs, your beliefs control your behavior and actions, and your actions deliver your results—failure, mediocrity, or resounding success.

It is like an iceberg.

What you and others see on the surface are your behaviors, actions, and the results you get. Under the surface are your *beliefs*. Deeper down, *thinking* controls your beliefs, and at the very

30 Adapted from "Emotional Intelligence," Wikipedia, accessed December 10, 2019, https://en.wikipedia.org/wiki/Emotional_intelligence.

deepest level, *emotions* influence your thinking. Much of the time, you are not aware of those deeper layers controlling the actions on the surface.

In 1983, Howard Gardner's *Frames of Mind: The Theory of Multiple Intelligences* introduced the idea that traditional types of intelligence measurement such as IQ fail to fully explain our mental abilities.[31] For example, IQ fails to measure the intellect of people gifted in music, art, or sport.

Gardner introduced the idea of *multiple* intelligences, which includes *intrapersonal intelligence* (the capacity to understand oneself and appreciate your own feelings, fears, and motivations) and *interpersonal intelligence* (the capacity to understand the intentions, motivations, and desires of other people).

However, the publication of Goleman's 1995 best-selling book *Emotional Intelligence: Why It Can Matter More Than IQ* saw the subject hit mainstream consciousness.[32] Later, in 1998, his book *Working with Emotional Intelligence* brought the subject to the attention of the business world. While theories have been developed to define and explain EQ, we have Goleman to thank for making the subject readily accessible to the business world.

In *Working with Emotional Intelligence*, Goleman logically argues that one might have a high level of EQ, but unless it is used effectively in the workplace, it remains ineffective, rather like a highly skilled sports person or gifted musician (for one reason or another) never using their talents. He defines the effective use of EQ in the workplace as an emotional *competence* or a *learned* capability based on EQ, resulting in outstanding performance at

---

31 Howard Gardner, *Frames of Mind: The Theory of Multiple Intelligences* (New York: Basic Books, 1983).

32 Daniel Goleman, *Emotional Intelligence: Why It Can Matter More Than IQ* (New York: Bantam Books, 1995).

work (I have emphasized the word *learned* because one aim of this module is to give you the knowledge to *learn* how to become a master of emotions in your business life).

The first time I read Goleman's book was another "light-bulb moment." It gave me clarity for why I shone in some areas of business management but failed miserably in others. As I related in the *Introduction*, I provided outstanding client service but would run a mile from conflict. I had great vision but was often paralyzed by fear of failure.

I could also look back on my dysfunctional business partnerships, at the strengths and weaknesses each of my partners brought "to the table," and how their lack of EQ in critical areas led to disharmony and unhappiness. However, I am not critical of my ex-partners. We all brought our individual strengths and weaknesses to the table, unaware of EQ and how to improve it, blind to the fact we could become competent in EQ or even develop a high EQ.

We were numbly flying in a fog without radar, waiting to crash.

EQ is not only important for you as a business owner or an entrepreneur but also for how you handle the emotions of your team and your clients.

As a young graduate and employee, I drove some distance to one of our client's businesses early one morning. To my dismay, I found that I had run out of some necessary equipment. I apologized to the client for the inconvenience, cursed myself, and drove all the way back to base for new supplies.

As I reached the storeroom, my boss at the time said, "Gee, that was quick!"

"Err, forgot some gear," I muttered.

Did he let me have an earful?! "What a waste of valuable time. What poor service—a top client, too!"

He went on and on, white with anger. To make matters worse, our receptionist was politely standing next to him, waiting to ask a question.

I was totally humiliated. Shaking and intimidated, I drove back to the client's property.

Now, I am not exaggerating when I say that my boss prematurely returned from an outcall the very next day.

"Gee, that was quick," I said, unable to help myself.

"Yeah, forgot something."

He coolly turned his back, got what he wanted, and drove off in a cloud of dust. I did not challenge him because of his senior position and, tellingly, because of my fear of conflict!

What was the level of my boss's EQ? He lacked self-control, he could not keep his temper in check, and I could not trust him. He failed to maintain his own standards of integrity. He also lacked empathy as he was unaware of my feelings, especially in front of a fellow team member.

I felt discouraged and disengaged. If he treated me that way, why should I work so hard for him?

We can also reflect on the strength of neuroplastic change when emotions are involved. Do you think this experience formed a neural connection or two in my brain? I can still remember where I was standing when he delivered his broadside and the details of his hypocrisy the next day. The negative brain cell network of the incident is firmly etched in my brain even today as I write about it.

> ***It is the Eleventh Key in action: Negative emotions instantly increase the strength and number of connections of, in this case, a new brain cell network.***

At the time, I knew that his behavior was not a path to amazing business success, but I did not realize how my own level of EQ also had a role in the incident, allowing him to get away with his behavior time and again. His EQ and my EQ were both involved in the power play—*and* on public display!

Incidents like the one I have just related—and millions of others involving our many different emotions—are played out in every business around the world every day. There are billions of positive and negative intra- and interpersonal interactions powering our businesses to success, allowing them to languish in mediocrity, or even plunging them into failure.

What stories can you tell about EQ (or lack thereof) in your workplace? You can now understand that EQ goes a long way to explain why some of those top performers in your classes at school or college subsequently failed in the business world and why some of the strugglers soared to success. Most of them will have had little or no knowledge of the concept of EQ. They used their own individual abilities (or inabilities) to manage their own (and others') emotions to navigate the complexities of their business lives. They went on to become successful or mediocre or to fail.

As *people* are responsible for managing and leading businesses, they often succeed or fail because of the level of EQ demonstrated "at the top"—not because of their own (or their team's) IQ, their products or services, or even changing market conditions!

How many small businesses remain stuck in mediocrity because their owners lack motivation? How many medium and large businesses fail because their executives are poor catalysts of change, failing to initiate or manage important changes in their industry? Motivation and change management are but two aspects of EQ.

We all have our own individual levels of EQ—our "baseline EQ." We all have our own strengths and weaknesses. Our emotions are with us the whole time; we depend on them for survival. We cannot help but be emotional beings. So, what are *your* emotional strengths and weaknesses? Now is the time to find out.

Using the following list, consider each one of these specific areas of EQ and reflect on your competence:

Self-awareness:	Conscious knowledge of your own character, feelings, and actions.
Empathy:	The understanding and sharing of the feelings of another person or other people.
Self-assessment:	Evaluating yourself or your actions, attitudes, behaviors, and performance.
Self-assurance:	Feeling confident in one's own abilities or character.
Self-reflection:	Serious thinking about your character and actions.
Self-management:	Taking responsibility for your own behavior and well-being.
Reliability:	Being trustworthy or performing consistently well.
Flexibility:	Sensing your willingness to change or compromise.
Innovation:	Creating and developing new ideas, methods, or products, etc.
Inspiration:	Filling others with the urge or ability to do or feel something, especially something creative.
Accomplishment:	Achieving something successful.
Integrity:	Being honest and having strong moral principles.

Enterprise:	Using ingenuity and resourcefulness to seize opportunity.
Confidence:	A feeling of self-assurance arising from an appreciation of one's own abilities or qualities.
Compassion:	Sympathy and concern for the sufferings or misfortunes of others.
Gratitude:	Feeling thankful; readiness to show appreciation for and return kindness.
Interpretation:	Explaining the meaning of something.
Mentoring and talent sharing:	Your ability to effectively advise or train someone, especially a younger colleague.
Emphasis on business:	Having an efficient, practical, and systematic approach to one's work or a task.
Service orientation:	Positioning, coordinating, and producing a product or service that meets your customers' needs.
Capitalizing on opportunities:	Taking the chance to gain advantage through creativity and diversity.
Judging the balance of team or office politics:	Feeling how an alliance can express itself and how well you demonstrate the ability, capacity, and strength to deal with power inequalities.
Leadership:	The ability to apply successful strategies to encourage the adoption of shared vision, mission, and values.
Interaction:	Communicating with someone or being directly involved with someone or something.
Conflict resolution:	The EQ required to resolve differences, avoid discord or conflict, or find informal or formal processes that two or more parties use to find a peaceful solution to their dispute.
Vision:	Thinking about or planning the future with imagination and wisdom.
Facilitation:	Making an action or process easy or easier or helping others to do so.
Developing relationships:	Cultivating influential connections.
Teamwork and support:	The EQ required to lead and support others in gaining common objectives.

As you read these competencies, you are probably thinking that, like me, you are strong in some areas but weak in others. Perhaps you had not considered that some issues were important in business. That is OKAY! In fact, there are so many emotions in our psyche that if we tried to focus on all of them at once, we would fail. Our efforts would be too little in any one area to make a real and practical difference.

It is also important to point out that self-assessment of our own competence in any one area has its weaknesses. We all lack the objectivity to truly assess our own emotional strengths and weaknesses. We cannot see ourselves as others do, and our opinion will be skewed one way or another, producing an overly positive or overly negative assessment of our own EQ.

There are two other ways to make an EQ assessment:

1. HAVE AN IN-HOUSE ASSESSMENT CARRIED OUT BY YOUR BOSSES, PEERS, AND SUBORDINATES

This is a so-called 360 assessment by "all those around you." The advantages of an in-house assessment by your bosses, peers, and subordinates are that it is relatively objective compared to self-assessment and it occurs in the most relevant environment—your very own workplace—and is done by the people who know you best.

It is also cheap, can be completed at any time, and can be repeated to measure personal growth and the effect of any training in one or more competencies. That being said, it should be performed anonymously in order to encourage freedom of opinion, with the objective of personal growth clearly stated.

Conversely, it can invite self-analysis, with no appropriate coaching guiding the process. Depending on the size of your organization and your position within it, you may not have bosses, peers, or subordinates, and therefore the analysis may lack sufficient perspective.

Despite these disadvantages, in-house assessments are still valuable because they contain the objectivity of other people's opinions. I have received two separate 360 evaluations on my own performance in my business, and they were extremely helpful in my personal development as a leader and manager.

The second evaluation was conducted after I had written this book, and it was significantly more favorable than the first. Why? I had wired my brain for a higher EQ!

You can download an example questionnaire as a framework for a 360 evaluation from the online course.

Get valuable feedback about your own EQ!

2. BE ASSESSED BY A PROFESSIONAL HUMAN RESOURCE FIRM USING VALIDATED ASSESSMENT TOOLS

Human resource (HR) companies also provide external assessment, training, and coaching in aspects of EQ, as applied to business. The advantages of this are that it is externally validated, robust, and objective. Used by accredited external facilitators, it can provide valuable coaching and training. Its disadvantages include price and a relative inaccessibility to small business owners, some entrepreneurs, and sole operators.

Now, use Step 16 in the online course to self-assess your EQ!

Identify your emotional strengths and weaknesses in Step 16. This is critical. We will return to Step 16 later in Unit 4 of the course.

So, given your Point A (where you are now with respect to your EQ), how do you neutralize your weaknesses so that they no longer hinder or block your success? How do you transform your strengths into superpowers to turbocharge your performance?

As mentioned in Chapter 1, we can go to personal development seminars or workshops, listen to gurus, or read personal development books, only to find nothing changes or perhaps gets worse—despite our new knowledge. That was my experience.

I was not alone in seeking help. In fact, in 2016, the personal development or self-improvement industry (i.e., the motivational programs and products that seek to improve us physically, mentally, financially, or spiritually) was valued at over USD$9 billion in the United States (US) alone.[33]

This market includes books, CDs, motivational speakers, public seminars, workshops, retreats, webinars, personal coaching, websites, apps, Internet courses, training organizations, and more. It is a huge market. So, how do gurus, coaches, teachers, and trainers make sure the improvements desired by their students justify such expenditure? Equally importantly, how do they make sure that the desired changes *endure*, i.e., we do not revert to our Point A?

WHAT THE GURUS *SHOULD* TEACH: THE THREE STAGES OF PERSONAL TRANSFORMATION

Gurus, coaches, teachers, and trainers should effect change in three stages:

1. By a range of strategies, they should identify the current physical, mental, financial, or spiritual status, position, or skill levels of their student(s), or "Point A"—where you

33 *The U.S. Market for Self-improvement Products and Services,* Marketdata Enterprises, 2017, https://www.marketresearch.com/Marketdata-Enterprises-Inc-v416/Self-improvement-Products-Services-11905582/.

are now. This process of identification should be highly personalized. We all differ in our needs and wants, so focus is important.

2. They should *transform* their students' status, position, or skill level to a new level of development or improvement, as is desired by the student. This is your "Point B"—where you want to be.
3. Finally—and this is the challenging bit—they should *embed* change so that improvement endures, ideally instilling a process by which *continuous* improvement is sustained beyond their direct influence. In other words, the student continues to grow.

This is a critical point! Given that EQ is at least *twice* as important as technical expertise (or IQ) for business success, how do we embed important positive changes into our personal development?

How *do* you become a Jedi Master of emotions, a Luke Skywalker of your business world?

Like Neo in *The Matrix*, how *do* we "download" new or improved thought patterns and emotions directly into our brain for when we need them most?

The answer to these important questions lies in the integration of the Three Stages of Personal Transformation—identification, transformation, and embedding—with The 12 Keys to Wire Your Brain for Success.

This is how you do it:

Stage 1: Identify! Select the critical strengths and weaknesses in your thought patterns and emotions you want to change

In Step 1 of the online course, you can identify your negative mindsets and self-limiting beliefs—the "baggage" blocking your amazing potential. Likewise, in Step 2, you can highlight the positive thought patterns and strengths that make you the amazing and unique person you are.

In Step 16, you can identify the EQ weakness you need to conquer and the EQ strengths you most want to convert to superpowers. Given that *you* are selecting them, your choices will be a highly personalized, exceptionally focused list of thought patterns and emotions you want to change.

Stage 2: Transform! Turn your weaknesses into strengths

This is the easy bit! Every negative and disempowering thought pattern or emotion has an opposing positive thought pattern or emotion. Transformation is simply the process of "flipping" the negative into the most positive and empowering version.

My most negative emotion was a feeling of anxiety or fear when faced with a conflict between others and myself. My response was total avoidance of the issue.

I "flipped" this into feelings of calmness and courage when resolving differences of opinion so that I could move the business forward to the mutual satisfaction of all parties.

My most negative thought pattern was the dread when presented with bad financial results, and I transmitted my thoughts to my team with my negative body language.

I flipped this into thoughts of opportunity about bad results, seeking to improve what we do. I inspired my team by leading that change and by looking forward to receiving the next set of financial results to monitor our progress.

And transform your strengths into superpowers

Conversely, transformation of your *positive* and empowering thought patterns or emotions is simply a process of improving or strengthening them, sometimes greatly so.

My most positive thought pattern was the enjoyment I gained when working with my clients and customers. What is the improvement? I now focus on creating massive, amazing value for my team, clients, and suppliers. A vastly better version! Value creation is now a superpower.

What is my most positive emotion in my business world? I always have feelings of excitement when I imagine the future direction of my business, including how it will look, sound, and feel in years to come. What is the improved version? I now share my vision with my team, allowing space for their ideas, making a more inclusive, inspiring, and improved working environment for all.

Now, be empowered with the all-important Step 17 in the online course!

It is one thing to be aware of new and highly desired emotions or thought patterns, but it is another to *embed* them in your mind. It is even more difficult to embed them so firmly that you can not only recall them under pressure but also *exemplify* them.

How do you escape your old habits and responses and develop an entirely new way of thinking and feeling? How do you embed or etch new beliefs or behaviors so that they become a fundamental part of your psyche—part of a new you?

We do not have Yoda to teach us emotional mastery.

We are not plugged into the matrix, where downloading new skills is easy.

As you may have guessed, the answer lies in The 12 Keys, which support and enhance one another, making progress easier and quicker than by using one Key alone. The following Keys are critical to the process:

> *The Second Key: Positive brain cell networks become stronger if we focus on positive outcomes.*
> *The Fourth Key: The strength and number of connections in a brain cell network increase with repetition.*
> *The Tenth Key: With every positive change, the brain takes the opportunity to weaken other connections.*
> *However, the most important Key in this process is the Eighth Key: Mental rehearsal, meditation, and mindfulness cause the same wiring in your brain as real-life experiences.*

So, let us enter the third stage of personal transformation.

Stage 3: Embed! Use mental rehearsal, meditation, or mindfulness to embed new and empowering thinking and emotions

Briefly returning to the computer analogy, imagine that mental rehearsal, meditation, and mindfulness are "software programs." Each one is available to embed the transformed versions of your thoughts and emotions; each is wiring and rewiring your brain with increasingly better brain cell networks; and each delivers those empowering thoughts and emotions.

All you need to do is "run" one of these programs when desired.

Unlike their computer equivalents, these software programs have some unique features and benefits. They are free; they need no power; upload speed is instantaneous—all you need to do is think of them; and they can be used at any time, in any place, and in any weather. Improvements come with practice. The benefits they deliver are good for you. If you wish, no one needs to know you are using them—these programs run in your mind. Privacy is assured!

The science is now in. Imaging studies of the brain have shown that repetitive *thought* causes wiring and rewiring, in some cases profoundly so.

It is the Fourth Key in action! Mental rehearsal, meditation, and mindfulness are all disciplines of repetitive thought, and science has likewise proven that neuroplastic change occurs when we rehearse, meditate, or practice mindfulness.

It is extraordinary that change is not limited to brain cell networks on a microscopic scale. Both the physical shape and size of areas within the brain increase or decrease with use or disuse. The brain is truly plastic!

So, Stage 3 of personal transformation is this: Use mental rehearsal, meditation, or mindfulness to embed the precise positive thoughts and emotions we have carefully identified. By letting them be, weaken the negative thoughts and emotions we do not want.

The first two stages—identification and transformation—are critical. We want to embed the *right* thought processes and emotional responses by design to suit your outcomes—*not* negative ones or random garbage!

The whole process is augmented and reinforced by using *all* 12 Keys.

Now, let us look at each of the "software" programs in detail and how you can "run" them in your mind.

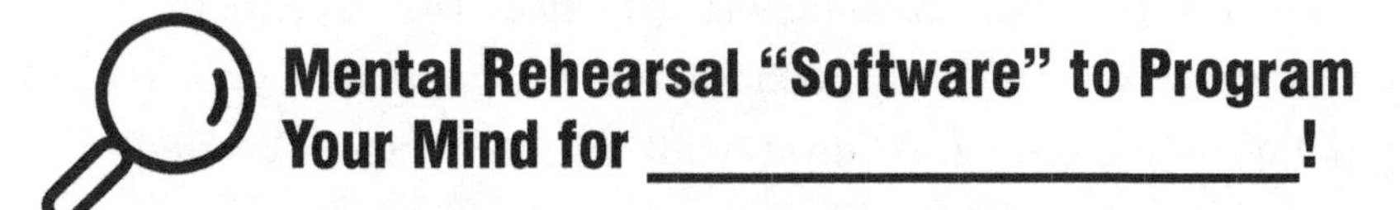

Mental Rehearsal "Software" to Program Your Mind for ____________________!

It would be inconceivable to think of an actor taking part in a stage show or movie without rehearsing. First, they receive a script and learn their lines. However, this only tells them what to say and when to say it. Then, they take part in rehearsals with other actors and the director, practicing time and again until their words and moves make a well-timed scene. Finally, in stage plays, they perform a full-dress rehearsal, coordinated with lighting, music, and sets. In movies, several takes are shot until the perfect scene is in the can, one which convinces us that all is reality.

Actors *hardwire* their brains to make the right moves at the right time, at the right speed, and in the right way. They will be totally unaware of the profound neuroplastic changes occurring in their brain as they go about their craft.

Athletes also mentally rehearse each step of their high-performance action until they achieve their optimum outcome.

When we train in our chosen vocation, we study for long hours, reaching the required level of expertise to receive a qualification. Upon graduating, real-life experience fine-tunes our skills until we achieve mastery. Both study and experience are forms of rehearsal, preparing us for high-level performance.

However, in our business lives, management and leadership require a high level of EQ for success, but we seldom rehearse for those roles. How rare are business courses on empathy, compassion, or courage? Yet these are some of the very skills required for a high EQ and for business success! Is it any wonder that some businesspeople often fail to rise above mediocrity? *They do not rehearse or practice for success*!

In my business, I noticed that if I had multiple demands placed on me, I became impatient and frustrated. I felt overwhelmed. I conveyed my impatience to my team in my body language and tone of voice. My head would drop, my shoulders would slump, and my eyes would roll.

Although I dealt with the demands, my non-verbal communication said: "I do not want to be disturbed!" I would even let out a low moan of frustration or impatience, which would convey: "Yeah, okay, what next?"

What messages were being sent to my team here, and how comfortable did they feel approaching me? My EQ in the skill of adaptability was poor.

To rewire my brain for adaptability, I physically sat down at my desk. I imagined writing an important report with a short deadline and then imagined having to answer an important phone call in which two team members asked me two different and important questions. On top of those priorities, I imagined being told I had an emergency case to deal with. Here is the important part of the sequence: I imagined remaining cool and calm, moving effortlessly from one task to the next.

I rehearsed this imaginary scene repeatedly until it became wired into my brain. Then, when the scene played out in real life, a member of my team asked: "How do you remain so calm?"

When I am tired, I still become impatient, but whereas I rated myself a 1/5 for adaptability, I now rate myself a 4/5.

A weakness had become a strength. I had not only improved my leadership skills but also made my workplace more enjoyable for my team. I had programmed my brain for adaptability, patience, and self-control using "mental rehearsal software." It works!

Now, go to Step 18 in the online course. Program your mind with mental rehearsal for _________! (You insert the outcomes you desire!)

Meditation "Software" to Program Your Mind for _________!

"All experience is preceded by mind, led by mind,
made by mind.
Speak or act with a peaceful mind, and happiness
follows like a never-departing shadow."
—From the Dhammapada, attributed to Buddha.

At first glance, it seems contradictory to use meditation—an ancient practice associated with many religions, and one that requires time, stillness, and a quiet environment—in the hustle, bustle, and pressure of modern business life.

It is an *invaluable* tool.

The topic and available resources are vast, and I invite you to explore it more deeply. I will confine my writing to the type of meditation I used to wire my brain for success. If you have not already practiced some form of meditation, there are many benefits.[34]

I use three sequential steps to meditate for success in my business and personal lives:

1. A slow and deep breathing pattern;
2. Full muscle relaxation; and
3. Focusing and refocusing on the subject of the mediation with a peaceful state of mind.

When my meditation is unsuccessful, I usually realize I have left one or more steps out.

So, to fully understand why we need the first two steps, go back 300,000 years to when primitive humans were leaving the jungles of Central Africa and adapting to life as hunters and gatherers on the open plains of East Africa. Out on the grasslands, man was both the hunter and the hunted: Trauma, high anxiety, and stress were facts of life.

Out hunting, the "fight-or-flight" hormone *adrenalin* would pump through early man's bloodstream, causing all the body changes we see in the fight of the hunter or in the flight of the hunted. His breathing would be shallow and rapid, delivering oxygen to the big muscle groups of his tense body.

In his environment, there would also be chronic, enduring stresses. Drought, food and water shortages, predators, and tribal and interpersonal conflict would be major stressors. These long-term factors would cause the stress hormone *cortisol* to surge in early man's bloodstream.

34 Emma Seppälä, "20 Scientific Reasons to Start Meditating Today," *Psychology Today*, September 11, 2013, https://www.psychologytoday.com/us/blog/feeling-it/201309/20-scientific-reasons-start-meditating-today.

Of course, there were times when conditions were favorable: shelter, plentiful food, the relative safety of company and darkness. Feel-good chemicals would dominate; his breathing would be deep and slow; and his muscles would be relaxed.

There was, *and still is*, a powerful interplay in our bodies between fight-or-flight and stress hormones and the feel-good chemicals. Our ancient ancestors needed this complex system to survive; it still exists unchanged within us, and we need to understand it to be successful.

Modern society, cultural expectations, pressures of work, personal relationships, and our very own personalities place demands upon us that can lead to anxiety, fear, and chronic stress—the very factors that cause fight-or-flight and stress hormones to rise in our bloodstream. They make us tense and interfere with our ability to relax, concentrate, and *successfully meditate.*

So, I "reverse engineer" the stress response to induce a peaceful state of mind by using a slow and deep breathing pattern together with muscle relaxation—the exact opposite of the rapid breathing and tense muscles of an anxious or stressed state. This reduces my adrenalin and cortisol levels and boosts my feel-good chemicals, helping me to achieve that peaceful state of mind and focus needed for effective meditation.

Now, to focus on each step in more detail:

1. USE A SLOW AND DEEP BREATHING PATTERN

This has long been associated with relaxation and a reduction in anxiety.[35] When you begin to meditate, it helps to find a quiet, darkened, private area, as free from external distraction as possible.

35 "Diaphragmatic Breathing," Wikipedia, accessed September 22, 2019, http://en.wikipedia.org/wiki/Diaphragmatic_breathing.

It may also help to play some relaxing music. However, when I am meditating, I do not play music with vocals because the words inevitably stir memories or other thoughts, taking me to "another place"; then, my mind is not focused on the subject of my meditation. Instead, I play music from the relaxation genre, with slow and soft sounds. It often makes use of the sounds of nature—the wind, the sound of waves, or birdsong. The music itself is not the object of meditation, but it takes your mind away from other sounds that could distract you, especially if you live in a city.

Next, I begin to breath slowly and deeply, gently allowing my breathing pattern to progressively slow down and deepen. I personally find that breathing at about six breaths each minute is appropriate. Your most effective rate will vary according to your age, health, and fitness.

This is a great first step for a meditation session. Try being stressed, anxious, or fearful when you are breathing very slowly and deeply!

2. ACHIEVE TOTAL MUSCLE RELAXATION

Once I am satisfied with my breathing pattern, I use a technique of progressive muscle relaxation.

Imagine you are lying down on a couch and I am sitting opposite you. I say to you, "Relax."

Likely, your shoulders would sag, and you might take a deep breath. Let us call that Level 1 relaxation.

Then, I say to you, "Close your eyes and take a slower, deeper breath." You might feel some tension drop from your body. Let us call that Level 2.

Compared with Levels 1 and 2, I will describe Level 10—*total muscle relaxation*. My introduction to this technique was, to say the least, unusual.

Long ago, a psychologist hypnotized me to see if I could help the police solve a murder investigation. Unfortunately, I was not able to help with their inquiries, but the experience stayed with me.

The psychologist led me to a quiet and darkened room in the police station and asked me to lie back on a comfortable couch. He told me he was going to place me in a very relaxed state; I could not enter this state unless I wanted to; and at the conclusion of the session, I would feel both happy and relaxed. I agreed to the process.

Initially, he shone a pencil-thin beam of light into my eyes and asked me to keep them open until he instructed me to close them. Then, he asked me to think of the little toe on my right foot, to wriggle it, and then relax it. At that stage, I really did not understand why he made this seemingly unrelated request, but I obeyed.

Taking his time, he asked me to do the same thing with the other toes on my foot, one by one, then my ankle, calf, knee, and my leg muscles. At this stage, my right leg felt heavy. The psychologist repeated the process with my left leg, toe by toe, then up the leg. Both legs then felt like lead.

Next, he focused on my fingers, wrists, elbows, arm muscles, and shoulder joints, still at a slow, deliberate pace. He moved to my back, abdominal and chest muscles, relaxing, relaxing, relaxing. I became aware that my breathing was very deep, slow, and calm.

He asked me to relax the muscles of my face, lips, cheeks, eyes, and forehead. At some point, the thin beam of light went out. I do not know if he turned the light out or if I simply could not keep my eyes open any longer.

Finally, he told me he was going to take me down from Level 1 to Level 10. He began to count, again with the same agonizing slowness: "one... two... three."

I am convinced that, at that point, the psychologist had his foot on the couch and was gently rocking it because it felt like I was in an old-fashioned lift, slowly descending into the depths of a dark but safe building.

"Four... five... six."

Never had I experienced such tranquil relaxation.

"Seven... eight... nine."

We reached Level 10—total muscle relaxation—a dark, beautiful place, still and safe. I had never known such a state of mind.

Back then, I decided to visit this place often. However, I never dreamed I would visit it to reduce the stresses and anxieties of running a business or that becoming skilled in relaxation was one step on a path to living a life I love.

If you would like to experience Level 10 relaxation, there is a range of options you can try:

- Experiment with the same process I experienced as described above. However, this may be difficult if you are alone. A sensitive partner may be able to take you through the same sequence.
- Purchase CDs to teach you progressive relaxation.
- Ask for a consultation with a qualified clinical psychologist trained in hypnotherapy.
- You may be able to reach similar levels of muscle relaxation with a massage given by a trained masseur.

- As I have done, float in a sensory deprivation tank!

The combination of a deep and slow breathing pattern coupled with total muscle relaxation creates a tranquil and peaceful mental state that permits successful meditation.

3. FOCUS AND REFOCUS ON THE SUBJECT OF YOUR MEDITATION WITH A PEACEFUL STATE OF MIND

One form of meditation, known as Zen meditation, teaches its followers to empty their minds and think of nothing. Teachers of this discipline advise that if thoughts arise, good or bad, just let them go—let them drift away.

There is a place for this type of meditation, for example, when you are stressed, anxious, or overwhelmed, or when you want "mental space" to solve problems or to be creative. I use Zen meditation in combination with Level 10 muscle relaxation to help me sleep.

However, to wire our brains for success, we focus intensely on developing *powerful, positive, and new* thought patterns or emotions, so thinking of nothing is counterproductive. First things first: Before we start meditating on business success, it is important to know *how* to meditate—how to find that peaceful state of mind and focus on our meditation subject. So, here are two meditations to practice.

Practice Meditation 1, using a candle as the subject of your focus

You need to select the *subject* of your meditation for your complete attention (i.e., your *point of focus and refocus*). I recommend a lit candle. There are reasons for my choice. It provides a target for your eyes focus on; you can look at a candle flame without hurting your eyes; and it adds soft light to a room. If you like, you can also use a scented candle to add a calming aroma.

You can choose any object you like for your meditation, real or imaginary. For example, real objects could include a rounded stone, an ornament, or a carving. Imaginary objects can also be used—a still pool of water with beautiful, deep-blue colors, clouds, or waves gently breaking on a sandy beach. Make it a scene you can easily imagine and enjoy.

It may also help to wear some loose-fitting, comfortable clothes that will not distract you. Again, play music from the relaxation genre.

I either sit on the floor with my legs extended out in front of me and my back resting on a soft chair, or I just sit in a comfortable chair. I allow my hands to rest by my side, or I fold them on my lap.

Meditation is often associated with the lotus position, in which the legs are folded with the feet resting on the inside of the knees, but I find this position uncomfortable, and it can interfere with my meditation.

If you like, imagine I am leading you in the following meditation, using the candle as the subject of your meditation.

> Breathe very slowly and very deeply. Relax all your muscles. Focus on the candle. Again, breathe very slowly and very deeply and relax. Just observe the candle flame. Watch it. See its color, its shape, its size, and its brightness.

Take in its scent. Breathe very slowly and very deeply. Allow your shoulders to drop. Watch the flame. Relax your muscles. Breathe more slowly and more deeply. See the flame. Relax.

The aim of this mediation is to see and think of nothing but the candle flame. If you think of something else, allow your thoughts to return to the candle.

Again, breathe slowly and deeply. Relax. Observe the candle. Relax your muscles. Breathe slowly and deeply. Observe the flame.

Meditation takes practice. At first, other thoughts compete for your attention, especially if you are anxious or stressed. Be patient and allow these other thoughts to drift away and return to the flame.

Here is a target: Focus your thoughts on the flame for the duration of at least one track of music. Say your music tracks last from two to ten minutes—that gives you some short and then longer targets. Start with short tracks and work your way up.

Success is focusing your thoughts on the candle for the duration of the soundtrack, without another thought but your candle flame. Build up until you can focus on the candle for the length of the longest music track without another thought entering your mind.

You should notice three things:

1. First, you will feel all the tension leave your body. You will feel wonderful relaxation.
2. Second, it is hard to feel stressed, anxious, or fearful.

3. Finally, it is a superbly peaceful and enjoyable feeling. While the candle is the *subject* of the meditation, the *object* is the amazing feelings you receive.

Practice Meditation 2, using a candle and your eyes closed

Start in the same way you did with the first meditation—with a lit candle. Again, I will lead you.

Look at the flame. Breathe very slowly and very deeply. Watch the flame. Relax your muscles. Breathe more slowly and more deeply. Observe the flame as before. Relax. Breathe deeply and slowly.

Now, close your eyes. You will still "see" the flame for a while but continue to see the flame in your "mind's eye"—in your imagination. Breathe very slowly and very deeply. In your mind, watch the flame. Think of its color, its shape, and its size. Relax. Allow your shoulders to drop. Breathe more slowly and more deeply. Imagine the flame. Smell the scent. Feel the warmth. Breathe slowly and deeply. Let your muscles relax... let them go. Relax. Think of the flame.

The aim of the two meditations is to lower the adrenalin and cortisol levels in your bloodstream and replace them with feel-good hormones. This will allow the stresses, anxieties, and fears of your life to fall away, letting nature take its course.

If these meditations work for you, it is my pleasure to give you this gift and the feelings that go with it.

Meditation enables you to practice finding that peaceful state of mind, and with it, the ability to focus and refocus on the subject of your meditation. They are a wonderful preparation for *wiring your brain with the specific thought patterns and emotions* YOU have identified as important.

In the *Introduction*, I related my experience of clinical anxiety and its devastating effects—the awful pressure across my chest as I made my way to visit clients; finding myself at 4.00 a.m., lying on the cold tiles of my kitchen floor in the fetal position, crying in desperation; or the energy-sapping distress of chronic insomnia.

This awful, heart-pounding, sickening feeling would bubble up inside me, sometimes for a minute, sometimes for a day, and sometimes without apparent cause—a raw, visceral fear of the future, of failure, of doom!

Then, at some point in this living nightmare, I remembered being hypnotized by the police psychologist long ago and his promise to me: At the conclusion of his session, I would feel both happy and relaxed. It was like a lifeline. This was what I craved—to be happy and relaxed!

So, the search began. I knew muscle relaxation was a key, so I purchased relaxation CDs, spending hours on my bedroom floor learning how to let each muscle go and relax. I made a little progress, becoming less anxious and more relaxed. Some of the CD titles suggested this music was also used for meditation. Meditation? At that time, I had not considered it could help.

My search changed direction and continued. I read book after book on meditation. Slowly, I developed the process I have outlined above, adding slow, deep breathing and the ability to focus and refocus on the subject of my meditation until… I was floating in a sea of calmness and relaxation. *What a relief!*

I had programmed my brain for calmness, relaxation, and even happiness using "meditation software." Like mental rehearsal, it worked amazingly well.

I am now going to jump in time and presume that you can meditate by successfully completing the candle meditations I have just described.

Now, utilize this skill in a very applied way to wire your brain for success. Return to Step 17 in the online course. Weaken the brain cell networks you have identified that control your negative thought patterns and emotions and strengthen those responsible for your empowering thought patterns and emotions—all with the power of meditation. This will *embed* your transformed thoughts and emotions.

The only limit to this exercise is your imagination!

Now go to Step 19 in the online course. Program Your Mind with Meditation for ________! (You insert the outcomes you desire!)

If progress is slow, here are some trouble-shooting tips.

First, develop the discipline of *practice*. Your ability to enter a peaceful state of mind and return your attention to refocus on your subject will improve with time. It is not necessarily easy at first, but it does get easier. With improvement, the duration of your meditation can become shorter—useful in a busy business environment—or longer, when you have time to devote to the discipline.

Second, if your attention wanders during meditation, as it is likely to do, develop the skill of returning to *focus and refocus* on the subject of your meditation.

Finally, learn to develop a *peaceful state of mind*—one that is not bouncing around from one thought to another or repeatedly troubled by pressures, stresses, worries, anxieties, or fears.

Enjoy!

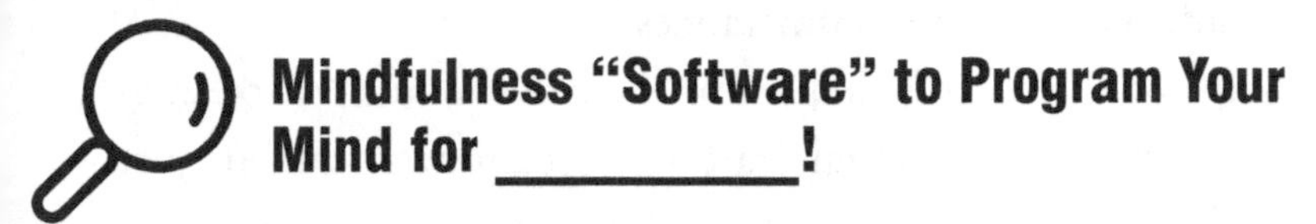

Mindfulness "Software" to Program Your Mind for __________!

We also wire our brains for success when we are being mindful. Since the turn of the millennium, the popularity of mindfulness has soared. It is now a USD$1 billion industry in the US alone.

More than 60,000 books on Amazon have a variation of "mindfulness" in their title, and many celebrities now practice some form of mindfulness, including Lady Gaga, Oprah Winfrey, and Goldie Hawn. Wall Street corporations like Starbucks and Proctor and Gamble, Silicon Valley giants Apple, Yahoo and Google, and even the US military have mindfulness programs. It has also been a subject of discussion and learning at the World Economic Forum in Davos, Switzerland.

So, what is mindfulness?

Wikipedia defines it as "the skill of bringing one's attention to whatever is happening in the present moment."[36] In other words, mindfulness is being in the "now" rather than thinking

36 See https://en.wikipedia.org/wiki/Mindfulness.

or worrying about the past or the future or allowing the mind to wander, daydream, or be overwhelmed, anxious or fearful.

The advantages of being in the now reached mainstream popularity with Eckhart Tolle's *New York Times* best-seller *The Power of Now*.[37]

For the purpose of wiring your brain for success, I am going to expand Wiki's definition to "the practice of bringing one's attention to whatever is happening in the present moment and intentionally and carefully selecting those thoughts or emotions that best suit our current circumstances."

This is not the purist's definition, i.e., that one tries to *avoid* producing a particular mental state (e.g., calmness) and to just notice each object that arises in one's thoughts without judgment on our ability to be mindful.

Consider this: There is little point in being in "the now" when we are at work, paralyzed by anxiety, instead of feeling productive calmness, or being fearful when we desire courage, or thinking we are worthless when we want to project personal power. Then, our mind is full of anxiety, fear, or low self-esteem.

> *As we now know from the Second Key, focusing on negative thinking and emotions only leads to more of the same. While there is much internal peace to be gained from being in the now with Zen-like nothingness in our minds, it does not achieve success in either our business or personal lives.*

37 Eckhart Tolle, *The Power of Now: A Guide to Spiritual Enlightenment* (California: Namaste Publishing, 1997).

So, let our minds be full—*mindful*—of precisely the thoughts and emotions we *want*, those that will achieve our goals and bring us closer to our wonderful vision of the future. Let our minds be full of calmness, courage, or self-worth when we need these thoughts and emotions the most in business and in life!

> *Let us draw upon the wonderful flip side of the Second Key—that positive brain cell networks become even stronger if we focus on positive outcomes and opportunities. This focus leads to even more positive thoughts and emotions—the type of thinking you want!*

Like mental rehearsal and meditation, mindfulness also needs practice.

> *Make use of the Fourth Key: Practice increases the number, strength, and durability of the positive networks in your brain—the outcomes you desire!*

Mindfulness is also important in the broader context of the business world because it results in superior employee well-being and more constructive work environments, as well as decreased levels of frustration, absenteeism, and chances of burnout. High levels of mindfulness are also associated with ethical decision-making, increased personal awareness, and improved emotional regulation.[38]

38 Adapted from https://en.wikipedia.org/wiki/Mindfulness.

Research also suggests that mindfulness training improves focus, attention, and the ability to work under stress—important competencies in business life.[39]

Multiple studies have also shown that mindfulness can be beneficial in treating symptoms of anxiety and depression—common conditions that are of enormous cost to the people who suffer from them *and* the businesses who employ them.[40, 41]

Not only did I use mindfulness to combat my anxiety, I also found it an exceptionally useful strategy to counter the chronic *busyness* that often negatively impacts our personal lives.

I love family gatherings! We have a big family, rich in character, individuality, and humor. However, during such occasions in the past, my thoughts often strayed back to my business—debt, team issues, challenging clients, and so on. I was not present. I was either in the past or in the future, fretting or worrying, missing out on the simple joy of being with my loved ones.

Once I became aware of my absence and the importance of being present, I used mindfulness "software" to program my brain to focus on enjoyment and fun at such times.

It made all the difference.

---

39 Katherine A. MacLean et al., "Intensive meditation training improves perceptual discrimination and sustained attention," *Psychological Science* 21, no. 6 (2010): 829–39, https://www.ncbi.nlm.nih.gov/pmc/articles/PMC3132583/.

40 Stefan G. Hofmann et al., "The effect of mindfulness-based therapy on anxiety and depression: A meta-analytic review," *Journal of Consulting and Clinical Psychology* 78, no. 2 (2010): 169–183, https://www.ncbi.nlm.nih.gov/pmc/articles/PMC2848393/.

41 Adapted from https://en.wikipedia.org/wiki/Mindfulness-based_stress_reduction.

This is such a seemingly simple example of the benefits of applied mindfulness, yet quality social connections are so important for our mental health and our relationships. Enhance those connections through better, more effective *brain cell connections*, leading to better feelings and emotions for all!

Now, go to Step 20 in the online course. Program Your Mind with Mindfulness for ________! (You insert the outcomes you desire!)

You will have beliefs and behaviors that do not serve you well. This is your chance to change them. You will have wonderful strengths. This is your chance to make them better. With better focus and freedom of thought, create new skills, too.

"Run" the "software" of mental rehearsal, meditation, and mindfulness *for your outcomes*. Wire and rewire your brain!

Like Neo in *The Matrix*, download specific thought patterns to acquire new skills or enhance existing ones.

Like Luke Skywalker under instruction from Yoda, amplify your wonderful positive emotions like empathy, gratitude, joy, and happiness to better control your negative emotions such as impatience, anger, and fear.

Let us be clear on what we are doing: We are powerfully altering our brain cell networks to enhance our performance in business and in life. This is based on scientific facts proven by world-leading neuroscientists and psychologists.

Once you have successfully rewired your most important thought patterns or emotions, you can select others to tackle. I advised you to start with two or three so that you focus on what is important to you and you do not feel overwhelmed.

There are some important points to make.

First, this process takes time. Mental rehearsal, meditation, and mindfulness are skills to be learned first and *then* applied. Be patient.

As the First Key states, hardwired brain cell networks can be difficult to undo, but once you have experienced the first sweet taste of success, these skills become quicker and easier to learn, and with time, hardwiring can be undone!

If you master these skills, you will even find new ways to use them. I invite your imagination into play!

Second, these skills are exceptionally powerful because they all utilize the Fourth Key: The strength, number, and durability of connections in a brain cell network of a specific thought process increase with repetition. Especially with respect to meditation, by achieving a peaceful state of mind, we create a mental environment that permits refocusing time and again. With all three skills, we can "hit the replay button" repeatedly, both during "sessions" or whenever you like. It is quick, easy, and powerful!

Third, a student once asked me which is best—mental rehearsal, meditation, or mindfulness? All three have their roles. To use the analogy of stage or movie acting, meditation is similar to an actor learning their lines in the peace and quiet of their home; mental rehearsal can be likened to a full dress rehearsal on stage; and mindfulness can be compared to an actor performing live, his or her *mind full* of the role to play, with all its nuances.

Fourth, in Chapter 1, I posed important questions. How do you create a feeling of optimism when you are feeling pessimistic? How do you feel joy in an environment of despair? How do you feel gratitude when you are feeling ungrateful about your current situation? How do you have feelings of abundance when scarcity surrounds you? How do you create happiness when you are sad, calm when you are anxious, and love when you feel fear or hatred?

> *Flip the Ninth Key around! It can be as easy to cause positive rewiring in your brain as it is to create negative change!*

For every negative emotion, there is an opposing positive emotion ready to be created, embedded, and made more durable.

Therein lie the answers to these vital questions! Flip each one of the negative emotions around to its opposing positive version: Pessimism? Why, optimism, of course! Despair? Hope and joy. Ungratefulness? Gratitude. Meanness and scarcity? Generosity. Sadness? Happiness. Anxiety? Calmness and tranquility. Fear? Courage to press on. Hate? The most powerful emotion of all: love! *This* is the process of transformation.

Each one of these positive emotions is ready to embed with the "software" of mental rehearsal, meditation, or mindfulness.

Fifth, in his blog "How to Hardwire Resilience into Your Brain," Dr. Rick Hanson lists five ways to further embed the subject of your focus, making the relevant brain cell connections stronger, more numerous, and more enduring—a process he calls *enrichment*:

- "Linger on a subject for longer—five, ten, or more seconds. Remember, brain cells that fire together, wire together. If your attention wanders, refocus back to your subject.
- Intensify it. Open to it and let it be big in your mind. Turn up the volume by breathing more fully or getting a little excited.
- Expand it. Notice other elements of the experience. For example, if you are having a useful thought, look for related sensations or emotions.
- Freshen it. The brain is a novelty detector, designed to learn from what's new or unexpected. Look for what's interesting or surprising about an experience. Imagine that you are having it for the very first time.
- Value it. We learn from what is personally relevant. Be aware of why the experience is important to you, why it matters, and how it could help you."[42]

Hanson adds that you should reward yourself when successful. Tune into whatever is pleasurable, reassuring, helpful, or hopeful about the experience. This will flag the experience as a "keeper" for long-term storage.

Reward is, of course, the Seventh Key to success.

I would add that you can also use all your senses to further enrich the subject of your focus. For example, if calmness is your subject, what images would you associate with a calm feeling? Still

---

42 Rick Hanson and Forrest Hanson, "How to Hardwire Resilience into Your Brain," *Greater Good Magazine*, March 27, 2018, https://greatergood.berkeley.edu/article/item/how_to_hardwire_resilience_into_ your_brain#.

water, a lake, or a pond? Perhaps a desert at night? What colors would bring about calming feelings? Blue or gray? The colors of a sunrise or sunset?

What smells or scents would you associate with calmness? Incense? Perfume? The smell of the ocean or beach? What sounds would you hear if you were calm? Waves lapping gently on the shore? A soft breeze whispering through trees? Music to chill out?

What sensations would you link to calmness? The feeling of silk? A massage? A cuddle from a loved one?

The possibilities are only limited by your imagination, but the intention is to magnify the subject in your mind, further embedding it by strengthening the associated brain cell networks.

So, while we utilize the Second, Fourth, Eighth, and Tenth Keys in this process, *all* 12 Keys can be used to create and embed new emotions or thought patterns. Every Key augments this process of *self-directed* neuroplasticity!

As I promised in the *Introduction*, when everything is integrated and holistic, harmonious and synergistic, wonderful things happen!

Finally, I have used the words "a new you" throughout this book.

You have this extraordinary skill inside you, to strengthen your existing brain cell networks, weaken others, or create new ones—wiring and rewiring at over 1,000,000 connections and disconnections every second. Imagine this: Your thoughts and emotions designed by you, your plastic brain molded by you, and your supercomputer brain programmed by you!

In this process of personal transformation, you will still "feel like the same person," but your thinking will be so much better—clearer, calmer, more focused, more logical, confident, and constructive. Your emotions are richer, more enjoyable, and more meaningful. You affect others more positively—notably, your investors, team, suppliers, clients, and your family and friends. However, you are still the authentic "you."

You will have created a mental environment in which the creation of value is effortless—one that allows more time and less stress, along with a clearer path to freedom, wealth, and happiness.

Can you sense the opportunity here? It is yours to seize! Can you sense the potential? It is your untapped potential to unleash! Can you sense a new role? It is YOU, starring in the movie that is your life!

Yes, a new you!

Now, as promised, a bonus!

My Emotional Mastery Formula

In Chapter 1, I wrote that the world is changing at an ever-increasing rate. The pace of change has never been as great as it is now, and the pace of today will be the new slow in the near future. As a businessperson, rapid change is your reality.

We also face increasing pressure from so many angles—financial, technical, interpersonal, regulatory, and IT—and we must make better decisions at an equally faster pace.

If you manage team members, you will know that they each bring their unique personalities to work each day, and some are challenging to manage. Inspiring, motivating, and in some cases placating your team members add another layer of pressure. Also having to deal with clients or customers is not necessarily a walk in the park. Some are very, *very* demanding.

Pressure, pressure, pressure!

How do you cope with rapid change and massive pressure while growing *and* flourishing in this challenging world? How do you soar above the challenges in your demanding life to reach unimagined heights?

You do not have time for rehearsal or meditation. In fact, meditation is a technique usually practiced by removing pressure, taking your time, and immersing yourself in deep thought—hardly a useful technique during the cut and thrust of modern business life.

How do you wire yourself for success on the move, programming your brain to quickly adapt to your environment? Despite the pressures and speed of modern business, how do you remain in control of your thought processes and emotions, thinking clearly and positively ahead, looking toward your desired outcomes?

How do we feel strong, positive, empathetic, or powerful *just when required* instead of weak, negative, insensitive, or powerless?

My Emotional Mastery Formula answers this question. Simply put, it is a short, positive, and powerful meditation to enter a desired state of mind—indeed, mindfulness. There are three stages.

1. Inhale slowly and deeply and relax

Start by taking one breath in, very slowly and deeply. While inhaling, perform a shortened version of total body relaxation. Obviously, if you are standing, you will focus on relaxing your neck, shoulders, arms, and chest muscles, but the effect will still be *calming* and enable you to concentrate and focus.

To an observer, it will just look like you have taken a deep breath to gather your thoughts—and in some ways you have. The slow, deep, and measured inhalation, relaxation, and pause give you time for choice. With choice, your response is measured and consciously considered, not pulled out from your brain when you are on autopilot, acting unconsciously or reactively.

As soon as you feel the calming effect of relaxation, and before you begin to exhale, deliberately pause to *collect all your power of intention and concentration.*

2. Focus strongly on the action or emotion you desire

If an action is required, choose the best version of it. Visualize it intensely and powerfully. See yourself taking the action decisively before proceeding.

If emotion is required, simultaneously choose the best emotion for that moment. Imagine it well up deep within you. Let it rise from within your heart and soul, where your feelings are generated.

3. Exhale, committing to your action or emotion, absolutely, purposefully, and completely

The entire cycle should take from three to ten seconds. Depending on your circumstance, one cycle may be enough. However, the real benefit of this ultra-short meditation is this: One cycle can follow another, with each cycle intensifying the focus on your action or on the depth of your emotions.

I once had to work with an extremely valuable member of my team. His technical work was amazing. He brought a high IQ to the workplace and was extremely popular with the rest of my team and my clients. However, his ego was enormous, and he dominated every discussion we had.

My opinions were dismissed, and my abilities downgraded. I began to feel inferior. Although I was his manager, I often felt as though he were my boss.

Ultimately, every time I needed to discuss an issue that may have been contentious, I felt fearful. My heart pounded, and my unsteady voice betrayed my emotions. Working with him was certainly not a pleasure.

I used my Emotional Mastery Formula with a feeling of courage as my focus, both regaining control of my feelings and establishing a normal and positive working relationship with him. I was not seeking superiority—just to gain mutual respect and an appropriate balance of power.

Using my formula empowered me to do just that.

We have all witnessed people under pressure take a deep breath and consider their actions. Now, with my Emotional Mastery Formula, we are adding power to this human behavior by seizing the moment, designing it to suit our needs, multiplying its power by repeating it, and having the advantage of practice.

You can use this technique in hundreds of everyday situations. For example, imagine starting an important presentation by using the Emotional Mastery Formula with courage or calmness as your focus. Imagine comforting a distressed colleague using empathy as your focus. Imagine performing a physical task that requires precision. Breathe in slowly and relax as you focus intensely on the task before you. As you exhale, commit to the task with accuracy.

Feel the power of your mind. Take control of your thoughts and emotions. Take control of your life!

Feel the wonderful benefits of my Emotional Mastery Formula in Step 21 of the online course!

This is the last step in *Get Wired for Success—The Course.* If you have completed every step in the course, well done! Fantastic effort!

If you wish to delve deeper into the science of meditation and mindfulness at an introductory level, I have provided links to blogs by Debbie Hampton, Steven Handel, Richard Paterson, Alice Walton, and Kristyna Zapletal in the reference list. A more comprehensive review of the science can be found in the *Handbook of Mindfulness: Theory, Research, and Practice* by Brown, Creswell, and Ryan.

A RED-HOT TIP

Self-awareness—that is, being attuned to and recognizing your own emotions and thought processes—is a key component of EQ.

You can improve this skill by meditating with self-awareness as the subject of your meditation. For example, you could design a meditation in which you require enhanced self-awareness in an important meeting or interview. Visualize yourself being acutely aware of your own thoughts and feelings.

Then, in a real meeting or interview, catch yourself at the very point of thinking or feeling negatively. Flip the thought or emotion around in your mind to its exact opposite. For example, if you catch yourself feeling very pessimistic for no good reason, flip the thought or emotion to one of high optimism. Then, quickly use the Emotional Mastery Formula to reinforce the positive thought or feeling.

In his article "Paying Attention," Dr. Rick Hanson writes the following:

> "In particular, because of what's called 'experience-dependent neuroplasticity,' whatever you hold in attention has a special power to change your brain. Attention is like a combination spotlight and vacuum cleaner: it illuminates what it rests upon and then sucks it into your brain—and yourself."[43]

Often, negative emotions or thoughts can tumble one after another, until you are caught in a cycle of negativity and poor results. Boredom, ruminations, despair, anxiety, or fear can build and threaten your whole day and well-being.

Be self-aware. Then, use my Emotional Mastery Formula to stop the negative cycle cold.

ANOTHER RED-HOT TIP

Sometimes, we frustrate ourselves. We fall short of the person we really want to be, and there are so many roles to play in our increasingly busy lives—loving husband or wife, father or mother, son or daughter, relative, friend, confidant, businessperson, leader, entrepreneur, marketing guru, financial wizard, technician, coach, or sportsperson—to name a few of the roles you play in your busy, busy life.

Is it any wonder we fall short of the goals and dreams we set for ourselves?

It is often our loved ones that bear the brunt of our frustrations as we painfully bear witness to the gap between our dreams and reality. The science of neuroplasticity, together with the powerful tools I have outlined in this course, provide you with the opportunity to play the many roles in your life with direction

43 Rick Hanson, "Paying Attention," https://www.rickhanson.net/pay-attention/

and purpose, with greater ease and confidence, and with less frustration, stress, or anxiety.

> *So, use the Sixth Key: Remember that models guide positive wiring and rewiring. In your mind's eye, design an ideal model of yourself. Create an "amazing you!" How would the amazing you see your world? How would you think and feel? How would you act?*

Now, imagine yourself moving effortlessly from one role to another… not just coping with life, but *creating* a wonderful life, powerfully and positively affecting those around you, too.

Mentally rehearse these big roles. Meditate on the person you really want to be, wiring your brain for greater success. Mindfully move through your day, programming your mind for your outcomes. Use the Emotional Mastery Formula to fine-tune your busy life on the run. Use the model of the ideal you as a framework for today, tomorrow, and your future.

Yes, be the *star* of the movie that is your life!

KEEP IT SIMPLE

- Emotional intelligence or *EQ* is twice as important as technical expertise or IQ for business success.
- Your EQ is your ability to recognize your own and other people's emotions, identify different emotions correctly, and use this information to guide thinking and behavior in yourself and others.
- In-house assessment of your EQ by your bosses, peers, and subordinates (the so-called 360 assessment) and

assessment by professional HR firms are more objective than personal review.

- The Three Stages of Personal Transformation are identification, transformation, and embedding.
- Identify the EQ strengths and weaknesses that you most want to change in Step 16 of the online course.
- Returning to Steps 1 and 2 in Unit 1, also identify the positive and negative thought processes you most want to change.
- Transform or "flip" the negative thoughts and emotions into positive versions, and make your strengths even stronger.
- *Use the Eighth Key to wire your brain for success. Use mental rehearsal, meditation, or mindfulness to create new brain cell networks, embedding positive emotions or thought patterns for when you most need them.* They are the "software" to program your mind for ________! (Fill in this space with the outcomes you desire!)
- Actors, athletes and, yes, businesspeople can use mental rehearsal to achieve a desired outcome.
- The three key components of successful meditation are (1) the ability to develop a peaceful state of mind; (2) the skill to return your attention to focus and refocus on the subject of your meditation; and (3) practice.
- I recommend three steps used in combination to meditate successfully: (1) adopt a slow and deep breathing pattern; (2) perform total muscle relaxation; and (3) the practice of meditation itself.
- Mindfulness is the practice of being in the "now." Practice bringing your attention to the present moment.

Intentionally and carefully select those thoughts or emotions that best suit your current circumstances.

- The Emotional Mastery Formula gives you a short form of meditation or mindfulness to use when you are stressed, under pressure, or time constrained.
- Self-awareness is a key skill in gaining high EQ.
- In your mind's eye, design an ideal model of yourself. Mentally rehearse your big roles in life. Meditate on this person you really want to be! Align your desired thoughts and emotions with mindfulness. Be the *star* of the movie that is your life!

TRANSFORM YOUR COMMUNITY

There is a new phrase in the English language: a thought leader—a person who leads others by their original thoughts. How relevant!

Design and embed the thought patterns and emotions you need to be a leader of thoughts in your business and in life. Wire your brain for inspired leadership, and then lead others in their thinking!

Be an example of beliefs, behaviors, and actions to your investors, team, suppliers, and clients.

Show them the way forward with your crystal-clear vision, motivate them with your compelling purpose, and lead them with your values.

Your community is begging for examples of exemplary beliefs, behaviors, attitudes, and actions. It needs clarity of thought. Be that leader of thoughts!

THIS IS ABOUT YOU

In Chapter 1, I explained that your brain is plastic. Once you know the *how*, choices multiply, and the paths to follow are more expansive. *You* can do it! Your paths *are* becoming more expansive!

If you have completed the 21 Steps, you will have a crystal-clear vision of the business life of your dreams and of you living a life you love. You will have a compelling mission to give you intrinsic motivation. You will know your life's true purpose. You will have values to point the way, the red-hot passion to inspire others, and now the "software" to master your thinking and emotions.

These processes and techniques will give you a new personal power—the power to create a wonderful life.

Old and negative brain cell networks that we identified in Step 1 become faded memories. Your strengths become even stronger, fine-tuned with practice. Exciting new opportunities begin to clamor for your attention, and the creation of value becomes effortless.

Like Neo in *The Matrix*, you can now download new brain cell networks to fundamentally, powerfully, positively, and passionately change the way you think, feel, and see your world. Program your brain for new beliefs, attitudes, behaviors, actions, and results! Become the new you!

Like Yoda and Luke, you begin to exert amazing control over your emotions, and through your example, you ignite the emotions of others within your sphere of influence. You will wire their brains for success! They will catch your focus (through your vision), your intensity (through your compelling purpose), and your integrity (through your values). They will catch onto what "you are about." They will become your followers, helping you to your vision.

They are gold! Through them, there are no limits! Reach for the sky!

Do not become the master of your destiny. You *are* the master of your destiny.

NEXT:

You have covered a great deal of information. Even with the sequence of the 21 Steps, it is reasonable for you to feel overwhelmed or discouraged, thus making no progress.

Chapter 4 may challenge some of your previous and perhaps tightly held perceptions. It is reasonable to be skeptical of, for example, the use of mental rehearsal, meditation, or mindfulness in your busy business life. Relevant questions may arise. If initial progress is imperceptibly slow or even negligible, how do you *know* if self-directed neuroplasticity—wiring and rewiring your brain for your desired outcomes—really works? How often should you check your progress? What time is required to see or feel a sustained change?

After all, positive feedback is important. Positive feedback is a reward for progress, and, as we know from *the Seventh Key, reward drives positive wiring*. We all want to feel rewarded.

So, in the *Conclusion*, I will set you a challenge and give you the Three Signs of Success.

CONCLUSION

"One can never consent to creep
when one feels an impulse to soar."
"No pessimist ever discovered the secrets of the stars or
sailed to an unchartered land or opened a new heaven to
the horizon of the spirit."
—HELEN KELLER, deaf and blind American lecturer, writer, and scholar, 1880–1968.

YOUR CHALLENGE: UNLEASH YOUR AMAZING POTENTIAL

Throughout *Get Wired for Success*, there are dark undertones in my story of fear of conflict and anxiety—my mind-numbing insomnia and a downward spiral into ill-heath were the result.

In the *Introduction*, I wrote that the seeds of this stress, breakdown, and failure were set long before I bought my business. They were set in my childhood and adolescence and as I grew into the adult I had become. Now, in the *Conclusion*, I will outline how those seeds were set.

Both in our childhood and adolescence, we are exquisitely sensitive to our environment. Adverse experiences can wire our brains for all manner of negative memories, beliefs, attitudes, and behaviors, plaguing and tormenting us in our adult lives—the "baggage" we carry around.

It was only when my mother died that I realized just how much I loved her and the debt I owed her. At her death, I cried out a howl of pain. Tears rolled down my cheeks as I looked at her body, now silent forever.

She had pointed me to science and business, to endeavor and discernment, and to values of quality and goodness. Her role in my life was massive.

However, growing up with her was far from easy. Mum expected me to be perfect. She was raised in extremely hard times, and because I was an only child, she had just one shot at vicariously living a life for which I suspect she had craved. I was going to make up for everything she had missed in her life.

My first memory as a child was of her repeatedly smacking my hands at the dinner table for holding my knife and fork the wrong way. During my years at elementary school, every weeknight consisted of spelling or math tests. Errors were cured by endless repetition. The only feedback was constant disapproval and correction of my many mistakes. I never heard a single word of praise—not one—so my confidence did not get off to a good start.

Enter stage left my intense desire to be perfect and my low self-esteem at failing to be so.

Mum and Dad bickered endlessly. When Mum hit menopause, their relationship slipped into open warfare, with her screaming endless frustrations at Dad. I can remember cowering in my bedroom, scared of the conflict, acutely embarrassed at what the neighbors must have thought of our family.

Enter my fear of conflict and why I avoided it as an adult.

By the time I entered secondary school, my confidence was shot. The environment in a boys-only private school can be brutal. Bullies pick out a victim easily—and that victim was me.

By my last two years in secondary school, the whole class (and even one teacher) would chant cruel and disgusting names at me. At breaks and lunchtime, with only one brave exception, not a single student would talk to me. I ate every lunch alone on a bench as the other boys played and socialized around me. I was totally isolated and insecure. I believed I was a complete failure and a misfit.

Enter my pessimism, anxiety, and fear of failure.

Social isolation on such a systematic scale is massively painful, devastating, and distressing. Many years later, those learned self-beliefs and subsequent fears threatened to derail my life, as I slid into an abyss of anxiety and insomnia, buried under $1.1 million of debt.

Here is the critical point: *They became distant, benign memories.* I learned that my brain is plastic. I wired it for success. I learned to escape my past and design a new future.

I conquered my self-imposed limitations, doubts, insecurities, and even my anxieties and fears. I became a better person. I leveraged my strengths and my creativity blossomed.

It is a rough rule of thumb that our genetics are responsible for about half of the person we become, and our environment is responsible for the other half. While we cannot do anything about our genes, and we cannot forget our deeply embedded memories, we *can* do a lot about the devastating effect our environment and experiences sometimes have on our thinking, emotions, psyche, and how we view our world.

The next critical point is that your brain is plastic, too!

In Chapter 1, I asked you these questions:

What are your negative mindsets and "baggage?" How do they limit your business life, personal life, and growth? Do you confuse them with reality?

How can you leave this baggage behind? How can you crush your mindsets? How can you alter this default setting, this imperfect status quo, this balance between positive and negative thoughts and emotions?

The two most important questions I can ask you are as follows: Just what is your plastic brain capable of? What are YOU capable of?

You will have your own stories that have shaped your life (for some of you, very negatively). This book was written for you!

Use this information to *wire for change*, to rewrite the scripts you receive, and to create new thoughts, exciting feelings, and bold new visions for the future for richly rewarding success!

Yes, this is my challenge to you: Now, unleash your amazing potential!

Wire your brain for *your* outcomes!

Create the business life of *your* dreams!

Live a life *you love*!

THE THREE SIGNS OF SUCCESS

Congratulations! You have covered a great deal of information.

However, as you tread down the path of life, wiring and rewiring your brain, if you find your initial progress to be slow or even negligible, how do you *know* if this applied knowledge of neuroscience and positive psychology really works for you? I have found three invaluable signs of success to guide me. I call them That Flourishing Feeling, The Upward Spiral, and Flow State.

The First Sign: That Flourishing Feeling

The field of psychology has, until recently, occupied itself with the miseries of the human mind: anxiety, depression, obsessions, addictions, disease, and so on. However, a new branch of the science has developed: *positive psychology*.

Psychologist Dr. Martin Seligman cofounded this new and exciting field and devoted his subsequent career to furthering the study of positive emotions, character traits, and workplaces. As such, he is widely regarded as the "father of the science of happiness."

Dr. Seligman is the author of such books as *Authentic Happiness*, *Learned Optimism* and *What You Can Change and What You Can't*. In his work, *Flourish*, he lists five key elements to human well-being: *engagement, meaning, positive emotions, positive relationships, and achievement.*[44]

The first real indication I had that I was making real progress was a feeling of well-being and contentment and of growth and development. Purpose, passion, and success started to infuse both my business and personal lives. I had more time and freedom to do what I wanted, and I was less stressed and anxious about the future. Yes, I was really *flourishing*.

Consider each one of Seligman's five key elements:

---

44 Martin Seligman and Jesse Boggs, *Flourish: A Visionary New Understanding of Happiness and Well-being* (New York: Simon and Schuster, 2011).

- Engagement: My crystal-clear vision meant I was fully engaged with my work.
- Meaning: Revealing my life's true purpose gave me meaning every day (and it still does today).
- Achievement: Each day brings me a step closer to my vision (and it still does). No matter how small the step, I feel a sense of achievement when my day is done, and I celebrate that feeling.
- Positive relationships and positive emotions: By wiring my brain for success, my relationships and emotions became progressively more positive.

I had incorporated each one of Seligman's key elements into my life. No wonder I was flourishing!

This is your first sign: Your well-being will improve. Your life becomes easier, better, and more enjoyable. We all know that chronic stress causes illness—even my physical health had improved!

The Second Sign: The Upward Spiral

In his book *The Seven Habits of Highly Effective People*, motivational author and speaker Steven Covey used the concept of an upward spiral to describe an ever-increasing level of personal development should one follow his seven habits.[45]

45 Stephen Covey, *The Seven Habits of Highly Effective People* (New York: Simon and Schuster, 1989, 2004).

I have felt similar progression when wiring my brain for success. Your starting point is your Point A, where you intentionally begin to wire and rewire, heading to your perfect future—your powerful vision (or Point B).

Initially, progress is slow—almost imperceptible. Then, as you increase your understanding of the process and its real power, your rate of progression rises, as described by the ever-increasing upward trajectory of the spiral. Further, your breadth of understanding widens, as described by the outward expansion of the spiral.

Your thinking and the way you see the world improve at an ever-increasing rate and with equally greater breadth—and here is the bonus: Your results get better in an upward spiral!

It is important at this stage to comment on "Point B." As soon as we put a tight definition on your Point B, we limit your upward progression. I zoomed past my first Point B long ago. Your Point B shifts as we continue to develop and improve.

It is also important to emphasize that you can only recognize the upward spiral by looking back and remembering what you were like at various points along your journey. Your beliefs are now more empowering. Your attitudes become increasingly more positive. Your behavior becomes consistent, inspiring, and exemplary. Your actions become completely aligned to your vision, compelling purpose, and values. They deliver the results you desire. You think and see the world in a whole new way—a new you, a bolder you, a better you, but still the *authentic* you.

In the *Introduction*, I wrote: Do you have *a feeling deep within you* that things could be so much better, that you have so much potential waiting to be unleashed, that there are limitations in your life—some of which you can identify and some that are perhaps holding you back, just below the surface of your consciousness—and you want to break free from all that is holding you back and achieve your true greatness?

I was writing to the authentic you. That inner person is unchanged.

When you experience the "upward spiral," your way of thinking will have changed—less negative thought patterns, stronger positive brain cell networks, no limitations, no fixed mindsets—only the exciting future!

The Third Sign: Flow State

In his groundbreaking book, *Flow: The Psychology of Happiness*, Mihaly Csikszentmihalyi writes:

> "We have all experienced times when, instead of being buffeted by anonymous forces, we do feel in control of our actions, masters of our own fate. On the rare occasions that

> it happens, we feel a sense of exhilaration, a deep sense of enjoyment that is long cherished and that becomes a landmark in memory for what life should be like.
>
> This is what we mean by *optimal experience*."[46]

He goes on to state that he

> "developed a theory of optimal experience based on the concept of flow—a state in which people are so involved in an activity that nothing else seems to matter; the experience itself is so enjoyable that people will do it even at great cost, for the sheer sake of doing it."[47]

To me, the greatest reward for wiring my brain for success is the percentage of time I am in "flow state," really enjoying an optimal experience. Thinking back, I might have spent 1–2% of my time like this. For most of my life, I was buffeted around by those "anonymous forces." I presumed that was just how life was meant to be.

Slowly, all the negatives began to wash away. Slowly, all the positives grew. Slowly, new skills emerged, and I began to flourish. Then, the pace of my development and understanding quickened in an upward spiral. Finally, my time spent in flow state expanded. First, it was 20%, and then it was 50%. Suddenly, everything seemed to fall into place, and most of my time is now spent in a state of continuous flow.

What does flow state feel like?

---

46 Mihaly Csikszentmihalyi, Flow: *The Psychology of Happiness* (New York: Harper and Row, 1990), 2

47 Csikszentmihalyi, Flow: *The Psychology of Happiness*, 4.

You move effortlessly from one activity to the next, totally absorbed, with each activity a step on the path to your vision, powered by your compelling purpose and steered by your values.

You move through your day as easily as water flows from one pool to the next—so enjoyable, relaxing, and rewarding. Other people call this activity work. I call it a joy.

Each day, you feel excitement at what awaits tomorrow, next week, next month, and next year. Your interactions with others become deeper, more meaningful, and more authentic. Your emotions soar.

You feel a deep sense of contentment. Fun and humor are abundant. You have time for others, creating a profoundly positive influence on them, making their lives better, too.

Time expands. There is no hurry, for although you have many more steps on your journey, each one is enriching and rewarding, to be savored for what it represents and what it means.

You are in control of your emotions and thoughts. You feel that nothing could be better, but tomorrow is still better, and the next day is even better still.

This is the real reward for wiring your brain for success. Welcome to a state of flow.

This is the Holy Grail you have been searching for. This is what life is about… how living should be. This is *loving* the life you live!

Those feelings deep within, some suppressed, some denied, are now free to burst out, bubbling up into your full consciousness. You sense your potential and the amazing individual you are. No one else can be you, and you have so much to give.

Here you are, on the cusp on something truly great, standing alone in the world, fully open to it, arms spread out, fingers extended, legs apart, fully balanced, head back, and eyes wide open. You are ready for what awaits.

The future awaits!

A new you. Wonderful! Magnificent! Amazing!

I wish you well.

Rod

Rod Irwin, 2020

ABOUT THE AUTHOR:

Dr. Rod Irwin is a scientist, technical writer, and businessperson with over forty years of experience working in and owning businesses.

His passion is to transform science into media that will educate, inspire, and entertain. He has written hundreds of newsletters and magazine articles, been a regular radio guest, and appeared on TV. Dr. Rod now draws on his experience to make a powerfully positive difference to the lives of businesspeople with *Get Wired for Success!*

Dr. Rod Irwin lives with his wife Celia and their dog Bollie in a rural regional city near Melbourne, Australia.

His interests outside science and business are travel, downhill skiing, and reading.

CONNECT WITH ME

Stay in touch to receive news, updates, and further publications. Keep wiring your brain for success!

On LinkedIn:

https://www.linkedin.com/in/rod-irwin-81364568/

On Facebook:

https://www.facebook.com/Get-Wired-For-Success-979342545460193/

On Instagram:

https://www.instagram.com/getwiredforsuccess/

YOU QUALIFY FOR A BIG DISCOUNT!

WHEN YOU PURCHASE *GET WIRED FOR SUCCESS—THE COURSE*

Thank you for reading *Get Wired for Success!* Now take the next step to lock in your new knowledge. Go to one of the following links below to receive a big discount on your purchase of Get Wired For Success—The Course.

If you purchased the ebook, please visit https://www.getwiredforsuccess.com/ebookdeal, and if you purchased the print book, please visit https://www.getwiredforsuccess.com/bookdeal.

In a few easy clicks, you will be watching the videos and completing The 21 Steps. Because you have already covered Lesson 1, you can dive straight into Lesson 2, Unit 1. Learn by seeing, listening, and *doing*. It is so powerful. Remember my guarantee:

You will be actively wiring your brain for greater success in business and life!

When you complete *Get Wired for Success—The Course,* I would love to hear from you. Tell me how things were for you prior to hearing about the course, the effect that the course has had on your business or personal life, and how things are for you now. E-mail me at feedback@getwiredforsuccess.com

REFERENCES

Bradberry, Travis. "Could Mindfulness Make You a Better Leader?" World Economic Forum. January 5, 2016. https://www.weforum.org/agenda/2016/01/could-mindfulness-make-you-a-better-leader/

Breuning, Loretta Graziano. *Habits of a Happy Brain.* Avon, MA: Adams Media, 2016.

Brown, Kirk Warren J., David Creswell, and Richard M. Ryan, eds. *Handbook of Mindfulness: Theory, Research, and Practice.* Reprint edition. New York: The Guilford Press, 2015.

Covey, Stephen. *The Seven Habits of Highly Effective People.* New York: Simon and Schuster, 1989, 2004.

Csikszentmihalyi, Mihaly. *Flow: The Psychology of Happiness.* New York: Harper and Row, 1990.

Davidson, Richard, and Begley Sharon. *The Emotional Life of Your Brain.* London: Hodder and Stoughton, 2012.

Gardner, Howard. *Frames of Mind: The Theory of Multiple Intelligences.* New York: Basic Books, 1983.

Goleman, Daniel. *Emotional Intelligence: Why It Can Matter More Than IQ.* New York: Bantam Books, 1995.

Goleman, Daniel. *Working with Emotional Intelligence*. New York: Bantam Books, 1998.

Goleman, Daniel, and Richard J. Davidson. *Altered Traits: Science Reveals How Meditation Changes Your Mind, Brain, and Body*. New York: Avery, 2017.

Hampton, Debbie. "How Mindfulness Changes the Brain." The Best Brain Possible. November 20, 2016. https://www.thebestbrainpossible.com/how-mindfulness-changes-your-brain/.

Handel, Steven. "Mindfulness is Self-Directed Neuroplasticity." The Emotion Machine. May 17, 2011. https://www.theemotionmachine.com/mindfulness-and-neuroplasticity/.

Hanson, Rick. "Paying Attention." January 26, 2020. https://www.rickhanson.net/pay-attention/.

Hanson, Rick, and Forrest Hanson. "How to Hardwire Resilience into Your Brain." *Greater Good Magazine*, March 27, 2018. https://greatergood.berkeley.edu/article/item/how _to_ hardwire_resilience_into_your_brain#.

Hofmann, Stefan G., Alice T. Sawyer, Ashley A. Witt, and Diana Oh. "The Effect of Mindfulness-Based Therapy on Anxiety and Depression: A Meta-Analytic Review." *Journal of Consulting and Clinical Psychology* 78, no. 2 (2010): 169–183. https://www.ncbi.nlm.nih.gov/pmc/ articles/PMC2848393/.

Jobs, Steve. "Steve Jobs' 2005 Stanford Commencement Address." YouTube Video, 15:04, March 7, 2008. https://youtu.be/UF8uR6Z6KLc.

Lucas, George, dir. *Star Wars: Episode IV—A New Hope*. 1977; California: Twentieth Century Fox.

MacLean, Katherine A., Emilio Ferrer, Stephen R. Aichele, David A. Bridwell, Anthony P. Zanesco, Tonya L. Jacobs, Brandon G. King, Erika L. Rosenberg, Baljinder K. Sahdra, Phillip R. Shaver, B Alan Wallace, George R. Mangun, and Clifford D.

Saron. "Intensive Meditation Training Improves Perceptual Discrimination and Sustained Attention." *Psychological Science* 21, no. 6 (2010): 829–39. https://www.ncbi.nlm.nih.gov/pmc/articles/PMC3132583/.

Merzenich, Michael. *Soft-Wired: How the New Science of Brain Plasticity Can Change Your Life*. San Francisco: Parnassus, 2013.

Paterson, Richard. "Meditation for Beginners: The Complete Guide." *Think Less and Grow Rich*, October 14, 2019. https://www.thinklessandgrowrich.com/meditation-for-beginners-the-complete-guide/.

Pink, Daniel H. *Drive: The Surprising Truth About What Motivates Us*. Edinburgh: Canongate, 2011.

Rock, David. *Quiet Leadership*. New York: HarperCollins, 2006.

Seligman, Martin and Jesse Boggs. *Flourish: A Visionary New Understanding of Happiness and Well-Being*. New York: Simon and Schuster, 2011.

Seppälä, Emma. "20 Scientific Reasons to Start Meditating Today." *Psychology Today*, September 11, 2013. https://www.psychologytoday.com/us/blog/feeling-it/201309/20-scientific-reasons-start-meditating-today.

Tolle, Eckhart. *The Power of Now: A Guide to Spiritual Enlightenment*. California: Namaste Publishing, 1997.

"The U.S. Market for Self-improvement Products and Services," Marketdata Enterprises, 2017. https://www.marketresearch.com/Marketdata-Enterprises-Inc-v416/Self-improvement-Products-Services-11905582/.

Wachowski, Lana, and Lilly Wachowski, dirs. *The Matrix*. 1999; California: Warner Bros and Roadshow Entertainment.

Walton, Alice. "Different Types of Meditation Change Different Areas of the Brain, Study Finds." *Forbes*, October 5,

2017. https://www.forbes.com/sites/alicegwalton/2017/10/05/different-types-of-meditation-change-the-brain-in-different-ways-study-finds/.

"Diaphragmatic Breathing." Wikipedia. Accessed September 22, 2019. http://en.wikipedia.org/wiki/ Diaphragmatic_breathing.

"Emotional Intelligence." Wikipedia. Accessed December 10, 2019. https://en.wikipedia.org/wiki/Emotional_intelligence.

"Financial Crisis of 2007–8." Wikipedia. Accessed January 19, 2020. https://en.wikipedia.org/wiki/Financial_crisis_of_2007%E2%80%9308#Timeline.

"Mindfulness." Wikipedia. Accessed January 15, 2020. https://en.wikipedia.org/wiki/Mindfulness

"Mindfulness-Based Stress Reduction." Wikipedia. Accessed January 20, 2020.

https://en.wikipedia.org/wiki/Mindfulness-based_stress_reduction.

"Motivation." Wikipedia. Accessed January 25, 2020. https://en.wikipedia.org/wiki/Motivation

"Motivation." Wiktionary. Accessed October 8, 2019. https://en.wiktionary.org/wiki/motivation.

Zapletal, Kristyna. "Neuroscience of Mindfulness: What Exactly Happens to Your Brain When You Meditate." *Medium*, May 19, 2017. https://medium.com/@kristynazdot/neuroscience-of-mindfulness-what-exactly-happens-to-your-brain-when-you-meditate-7d1ca47d9fca

CPSIA information can be obtained
at www.ICGtesting.com
Printed in the USA
JSHW031227091220
10133JS00001B/34

9 781631 951145